"Joe Trippi is a classic American po... ...or, and a half-crazed competitor—an ... ...*vo-lution Will Not Be Televised* is not ... ...ous stories and strange characters, but ... ...ual for how to run a twenty-first-century political campaign. This is a book that will endure—about a groundbreaking campaign and the unforgettable political wizard who managed it."
—Joe Klein, author of *Primary Colors*

"Joe Trippi was one of the few people to recognize the power of what we were building in those first early, lonely days of the blogging revolution. He was certainly the first to try and harness that energy for a campaign. The current Democratic successes can be attributed, in large part, to Joe Trippi's embrace of revolutionary technologies that opened up politics to a new generation of Netroots activists."
—Jerome Armstrong and Markos Moulitsas Zúniga,
authors of *Crashing the Gate: Netroots, Grassroots,
and the Rise of People-Powered Politics*

"Joe Trippi gives us the ultimate inside story—and quite a story it is!—of the first presidential campaign to grasp the power of the Internet to transform politics and our democracy, Joe Trippi captures the passion of the Dean campaign as it reinvented politics for the twenty-first century."
—David Weinberger, coauthor of
The Cluetrain Manifesto: The End of Business as Usual

"Joe Trippi was part of the Internet team for Howard Dean, a group of people who changed American politics by involving millions of ordinary people in politics. He and many of the same people are now changing American history by successfully using the Net to unite many more millions in a manner that will elect the next president and revitalize our democracy. *The Revolution Will Not Be Televised*, particularly the new material, describes how these efforts are creating real, effective change in America, even as you read this."
—Craig Newmark, founder of Craigslist.org

"Joe Trippi wasn't just present at the birth of the first Internet-savvy presidential campaign, he was the midwife of the new politics we see emerging all around us today."

—Howard Rheingold, author of *Smart Mobs:*
The Next Social Revolution

"Joe Trippi is an eccentric pioneer and closet idealist, reckless enough to believe in the 'revolution' he saw coming in presidential politics. Back in 2004, he sounded like a crackpot inventor building a rocket ship in his garage. Now [he's] recognized as an authentic prophet, his vision dramatically confirmed by the 2008 presidential campaign. . . . Listen to this fun-loving pioneer explain how it happened."

—William Greider, author of *Who Will Tell the People*
and national affairs correspondent at *The Nation*

ABOUT THE AUTHOR

JOE TRIPPI was a senior campaign adviser for Senator John Edwards, working with Edwards to use the political energy of the Internet to fuel his 2008 presidential campaign. After working on the presidential campaigns of Senator Edward M. Kennedy (1980), Walter Mondale (1984), Gary Hart (1988), and Richard Gephardt (1988), he made national headlines in 2004 with his innovative Internet-based campaign for Governor Howard Dean. Trippi has worked on more than one hundred campaigns and has helped elect city attorneys, governors, mayors, and senators. He has worked internationally on campaigns in the United Kingdom, Italy, Greece, and Africa. Throughout his career Trippi has worked with a number of high-tech companies, including Wave Systems, Progeny Linux Systems, and IBM. He heads the multimedia consulting firm Trippi and Associates. Trippi is an on-air election analyst for CBS News and regularly appears on *Face the Nation*. He has been profiled in *GQ*, *Fast Company*, *The New Republic*, and the *New York Times Magazine*. The father of three, he lives with his wife, Kathleen Lash, on the eastern shore of Maryland.

THE
REVOLUTION
WILL NOT BE
TELEVISED

THE
REVOLUTION
WILL NOT BE
TELEVISED

DEMOCRACY, THE INTERNET, AND
THE OVERTHROW OF EVERYTHING

JOE TRIPPI

HARPER

NEW YORK · LONDON · TORONTO · SYDNEY

HARPER

A hardcover edition of this book was published in 2004 by HarperCollins Publishers.

HarperCollins books may be purchased for educational, business, or sales promotional use. For information, please e-mail the Special Markets Department at SPsales@harpercollins.com.

First Regan paperback published 2005.
First Harper paperback published 2008.

Designed by Publications Development Company of Texas

Library of Congress Cataloguing-in-Publication Data is available upon request.

ISBN 978-0-06-156107-8

14 15 RRD 10 9 8 7 6 5 4

To the six hundred thousand people of Dean for America
who relit the flame of participatory democracy
And to the flame of my life, Kathy

IN MEMORIAM

*Marc Cobb, my high school track teammate, who changed
my life by showing me the way to college.*

*Bill Warner, who always made me feel at home
when he was in the room.*

*Hunter Allen, one of the youngest and most beloved
Dean pioneers of 2004.*

God forbid we should ever be twenty years without such a rebellion . . . what country can preserve its liberties, if its rulers are not warned from time to time that this people preserve the spirit of resistence?
—Thomas Jefferson

Send lawyers guns and money
The shit has hit the fan
—Warren Zevon

CONTENTS

PART III
SEIZING POWER IN THE INTERNET AGE

AUTHOR'S NOTE

It's the spring of 2008, and I'm sitting here sick to my damn stomach. I'm envious and I'm exhausted and I'm in the last place a lifelong pol ever wants to find himself—on the sidelines, run out of the 2008 race for president by a campaign that took my own book and read it back to me "with meaning."

Let's just say Barack Obama got it.

I don't know exactly how things will play out between now and November. But in a lot of ways I don't think it matters. Maybe politics, like history, is a story eventually told by the winner, but in the annals of campaigning, what Barack Obama and his campaign staff were able to achieve in 2008 qualifies as yet another quantum leap in campaigning, in the use of the Internet, and in our democratic history—no matter what the final numbers say.

And as the strategist who first came up with and used many of the theories and ideas and technologies that the Obama camp used to further transform American politics on the eve of the 2008 election, I am proud and thrilled and, yes, more than a little jealous, like a biplane aviator watching astonauts walking toward the launch pad. Because at this moment, in the summer of 2008, Barack Obama and his campaign look to all the world to be performing the political equivalent of landing a man on the moon and returning him safely to earth. Theirs has been only the second bottom-up campaign in modern history, the second presidential campaign to get it. And it might just go all the way to the White House, with the American people providing the rocket fuel. There were few doubters left; this time, or maybe the next, the new way will get us there.

If Obama for America is the *Apollo 11* of a new politics, then what you're about to read in this book—the story of Dean for America,

Howard Dean's campaign of 2004, and those inspiring pioneers who worked with me on it—is like the Wright Brothers at Kitty Hawk. This is the story of the beginning of it all: of the first intrepid, seat-of-the-pants, spit-and-chewing-gum attempt to build a contraption that would eventually plug people back into their democracy, so that, sooner than any of us could have dreamed, we would fly.

Joe Trippi
Wittman, Maryland
May 2008

INTRODUCTION

December 2003

MY GUY IS about to crash and burn in an Iowa cornfield.

I can feel it. I have a sense about these things, especially in Iowa. I have a kind of clairvoyance in Iowa. I can smell death in Iowa.

While the candidate smiles and some of the staffers daydream about White House posts, our campaign has grown sick with all the symptoms of the old politics: infighting and petty jealousy among the campaign staff, gaffes by the candidate, cannibalistic ads by the other Democratic contenders—all of it beneath the steady eyes of the scavenger political press, always on the lookout for stray hunks of flesh.

No, we're going down.

And the worst part is this: There is nothing more I can do. After months of scraping and cajoling and pleading just to get the plane down the runway, now that we're finally aloft—and the rest of the crew is celebrating—I look out the window and the wings are coming off.

And I'm the only one who sees it.

I desperately want out.

Every fiber of my being is telling me to get out.

But I can't.

For the better part of a year, I have been the one person inside Howard Dean's presidential campaign saying that we could actually *win*. Back when I signed on as campaign manager, back when we had seven people on staff, $100,000 in the bank, and only four hundred thirty-two known supporters, back when you answered the phones yourself or they just kept ringing, back when Howard Dean was little more than an asterisk, the last name on a long list of Democratic presidential candidates, I was the one looking people in the eye and telling them: *Look, we're gonna win this frickin' thing.*

Now, here it is the end of 2003, and we're actually on top, ahead in the polls, in the process of raking in more than $50 million, $15.8 million in this fund-raising quarter alone—a record—most of it from small donations of $100 or less. And whose fund-raising record are we beating? Our own! From the quarter before. We have an army of almost 600,000 fired-up supporters, not just a bunch of chicken-dinner donors, but activists, *believers,* people who have never been politically involved before and who are now living and breathing this campaign. Through them, we have tapped into a whole new vein of democracy and proven the Internet as a vibrant political tool. Now everyone is paying attention. The labor unions are beginning to endorse us. Al Gore has endorsed us. The media that we had to beg for coverage a few months ago has all but coronated Howard Dean as the Democratic nominee. We got the covers of *Time* and *Newsweek.* We *are* the story. And finally the other people in the campaign are beginning to mumble what I've been screaming for a year: *Hey, we're gonna win this frickin' thing.*

Only I don't believe it anymore.

The Iowa caucuses are a little more than a month away and we are bleeding. Our momentum is gone. Our message is getting lost. We're spending all our time and energy deflecting attacks from other campaigns. Our guy has become an unmitigated disaster on the road. The unscripted candor that served him when he was the longest shot is now being played like a sort of political Tourette's. The press continually mangles the context of what he says, amping up his words in their own cynical version of "Twist and Shout." We've got no adults with him on the road—no seasoned political people—and so, naturally, he's gaffing his way across Iowa.

The young Dean staffers—all energy and idealism—have no idea what's about to happen. For most of them, this is their first presidential and they don't realize that the only thing longer than the hours are the odds of winning. Some of them—the really crazy ones—have caught the bug and might work a second presidential. There could even be the odd addict or death-wisher among them who might someday forget how hard this was and work a third.

This is my seventh.

And I can see it coming apart. I can see that we've gone to the lead too soon, that the other candidates are bearing down on us. I know what hell there is to pay when an insurgent catches the mainstream party leaders off guard. I can practically hear the guns swinging around, the sights settling on our back. I've worked too many caucuses in Iowa to not immediately recognize the signs of this thing: the squabbling, the spending, the negative ads, the constant press scrutiny. I can see all of it beginning to take its toll.

Most of all, I can see that we just weren't ready. Not for this.

Before Howard Dean launched his presidential campaign, he made the dubious decision to seal many of his records as governor of Vermont for a decade—saying that he didn't want "anything embarrassing appearing in the papers at a critical time in any future endeavor." Well, it's a critical time now, and his decision has come back to bite us in the ass, this candidate who promised a new, open style of democracy hiding more than eleven years' worth of memos and files from the only major office he's ever held.

So here we are, in early December 2003, and the senior staff has decided to meet with the governor to plead our case for releasing the records. About fifteen of us have gathered in the long conference room on the third floor of a stale office building in South Burlington, Vermont—where this rebel campaign had its unlikely rise. We explain that everything is about to hit critical mass, and that we are under a new kind of pressure here. He is now the frontrunner—everything he does and says will attract new scrutiny—and he can't say out of one side of his mouth that he wants to clean up politics, while out of the other side say that his own records are off-limits for a frickin' decade.

We tell him that it's starting to show up in the polls. We can survive a lot of things, but we can't survive having people see him as just another double-talking politician. The Dean for America campaign is the antithesis of that . . . a grassroots, reform candidacy breaking all the old rules and making people believe in politics again.

"You've got to release the records, Governor."

His eyes are set, and his open face is pulled back defensively into that tree-trunk neck. "But there's nothing in there."

"If there's nothing in there, then we should release them."

"But there's nothing in there."

"That's why we have to release them."

"But why should we release them when there's nothing in there?"

We go around in circles like this until Governor Dean—whose running mate could have been stubbornness—ends the debate by saying he's done talking about it. "I would rather withdraw from the race than release those records."

We're all quiet. The frontrunner in the 2003 Democratic presidential campaign is threatening to quit, while he still has the lead. The meeting ends, Governor Dean nods in my direction and chokes out the words, "Follow me, Joe."

I try to keep up, but he's striding down the hallway toward my office, and I'm straggling fifteen feet behind him, reassuring staffers as I move down the hall.

My office is in the corner of the third floor, a long narrow gash of a room—a crash site of paper, CD cases, and empty Diet Pepsi cans. Howard Dean is standing against the wall, his back to me. He's shaking.

"You made this too easy," he manages to say.

"What?" I ask.

"This. I never thought it would go this far. I was going to raise my profile, raise health care as an issue, shake up the Democratic Party. Help change the country. But I never thought *this* would happen. Don't you understand?" He turns and faces me. "I never thought I could actually win. I wanted to . . . but I never really thought it could happen."[1]

SURGE OF POWER

I've spent my life moving from one election to the next, living out of suitcases, motel rooms and rental cars, sleeping on couches, knocking on doors, leafleting neighborhoods, writing ads and speeches and snappy debate comebacks trying to get a succession of Democrats elected to every office you can imagine—from city attorneys to U.S. senators, from mayors to several unsuccessful runs at electing the president of the United States.

In twenty-eight years, I would guess that I worked on more than a hundred campaigns. I won more than my share, some that I shouldn't have, and lost some that, to this day, still break my heart. I nearly killed myself for some of these people, a few of whom seemed born to lead, some of whom I would've had trouble punching the hole for myself.

Occasionally, I worked for a candidate who tapped into the original well of idealism that first got me into politics—an eleven-year-old

---

[1] Slightly different versions of this story have appeared in various media reports quoting Howard as saying he didn't *want* to be president. It's worth noting that only Howard Dean and I were in the room that day.

boy watching on TV as Bobby Kennedy walked off the stage toward his assassination, a gauntlet of hands, black and brown and white, reaching out to him, for more than just his plan to erase poverty or to end the Vietnam War. Hands that seemed to reach out for some kind of deliverance.

Like a lot of sick twists who practice politics as a career, before 2003, elections for me were always about the candidate. I would do anything for the candidate. I would work to the point of exhaustion. I would use every tool at my disposal. I would write television attack ads that made the opponent look like a polluting, Medicare-hating, bribe-taking sociopath who spooned every night with fat-cat businessmen and convicted murderers. I would do anything for the people I was trying to elect. My loyalty to them was everything.

And then this thing happened.

When Howard Dean's bid for the presidency finally did its crash-and-burn (not in an Iowa corn field, as it turned out, but onstage in an Iowa ballroom), a cynical, middle-aged campaign consultant who thought he'd seen it all, who thought he knew it all, an old pro who'd made it his life's work to win elections at all costs, learned the most profound and unexpected lesson of his life.

This time, it wasn't about the candidate at all. It was about the people. This was never about *him*. It was about them.

An amazing thing happened in the presidential contest of 2004: For the first time in my life, maybe the first time in history, a candidate lost but *his campaign won*.

When Governor Dean stood in my office and admitted that even *he* hadn't expected to be thrust into the lead for the Democratic presidential nomination, he was saying what I'd known for months. That this was bigger than him. Certainly bigger than me. Bigger than the Democratic Party. Bigger even than determining who ran against George W. Bush in the general election.

This was nothing less than the first shot in America's second revolution, nothing less than the people taking the first step to reclaiming a system that had long ago forgotten they existed. This was democracy bubbling to the surface, flooding the landscape, and raising all of us—an obscure Northeastern governor, his inexperienced supporters, and a handful of old political warhorses—along with it.

The Dean for America campaign arrived at just the right moment—a pivotal point in our political history, when forty years of a corrupt system had reduced politics to its basest elements—the race to raise money from one-quarter of one percent of the wealthiest Americans and corporate donors in exchange for dictating the policy of the country. Then, the side with the most money simply bought the most television ads to manipulate the most people—while instant polling, focus groups, and message testing refined the struggle to a few swing voters in a few key districts in a few key states, blurring any significant differences between the monolithic parties and destroying honest debate about issues like health care and the war in Iraq. Until every candidate sounded *exactly* the same, and a member of either party could proudly stand up and proclaim that his party had passed a Patients' Bill of Rights—an utterly meaningless bill that, incidentally, *didn't provide health care for one single American.*

If there is a playbook for this type of checkbook, top-down, cynical politics and governing, it was being written by George W. Bush's administration. Simply tell the voters that you're going to be compassionate, and then turn over the keys to the rich guys who wrote the checks. Hand the economy to the special interests. Turn the environment over to the oil companies. Wage war for the people who wrote the checks.

Against this backdrop of *transactional politics*, campaigns have become more vicious, more media savvy, more technologically advanced, more expensive and intensive, longer, bigger, and stronger in every way except one.

Somewhere along the line, they lost the voters.

As television transformed political campaigns, people began viewing elections as no different than any other product someone was trying to sell them—a new Chrysler, a new bacon-Monterey cheeseburger, a strapless pair of shoes. So they channel surfed. They tuned out. When the networks call elections before voters have even been to the polls, when they turn our political system into just another TV show (and not a very good one at that, something between the World Wrestling Federation and *The Real World*) all they do is encourage people to turn the channel.

So that's what we did. We turned the channel.

From that seminal moment when I watched Robert Kennedy declare victory and then turn and walk toward his death in 1968, until now, the involvement of Americans in all levels of politics has fallen precipitously.

I'm not talking just about the decline in the number of voters in the presidential election (which fell from 62.8 percent in 1960 to below 50 percent in recent years). The percentage of people who worked for a political party also plummeted, by about 42 percent in the past 30 years. The number who served on a committee for some local organization fell by 39 percent. Thirty-five percent fewer attended public meetings. Thirty-four percent fewer attended a political rally or speech.[2]

Across the board, Americans—made hopeless by a hope-killing process—have been leaving politics in droves.

I should know. I was one of them.

While I kept working in politics throughout the 1990s—mostly on TV ads for Senate and House candidates—I eased away from managing campaigns and began pursuing my other passion, technology. I worked for several computer and Internet companies, innovative, risk-taking twenty-first century businesses that threw away the old templates and began looking for new ways to do things.

Being a political junkie at heart, by the late 1990s I began daydreaming about a campaign that would be run the way these revolutionary companies were being run—not from the top down, with a $200 million TV ad budget and a detached board of directors, but from below: A campaign run by the people.

And that's when Howard Dean came along, an underdog so far out of the race we had no choice but to test this strategy, blending my two passions, bringing to the political world the things I'd learned in the technological world, taking democracy to the last place where democracy stood a chance.

The Internet.

I'd be lying if I said that when I made the first inquiry about using an obscure web site called MeetUp.com to link Howard Dean supporters together from around the country, that I knew in a year we'd have 600,000 people passionately committed to our cause. That these people would raise up the one candidate who actually seemed to have convictions, who rejected the old politics, who took the people seriously by engaging them and empowering them in the one place where they could meet him, the one

---

[2] Robert D. Putnam, *Bowling Alone: The Collapse and Revival of American Community* (New York: Simon & Schuster, 2000).

place where the ubiquitous presence of television couldn't distort his message—on Internet bulletin boards and web sites, chat rooms and web logs.

Certainly, I had known that politics would eventually come to this point, just like every other aspect of our society will eventually come to this point. I'd seen for years that the ingredients were there for overthrowing a decaying political system and replacing it with something responsive and revolutionary.

But I'd also be lying if I said that Howard Dean was the only person in my office that day stunned by the sudden power surge of Americans banding together to take back a system that had failed them miserably.

A DOT-COM MIRACLE

Everyone knows by now how the Dean campaign ended, in a looped tape of seemingly misplaced, eleventh-hour enthusiasm ("And Oklahoma! And Arizona! And North Dakota!"). Challenged by something they didn't create, couldn't control, and never understood, the networks and news media flexed their atrophying muscles and repeated that clip over and over, as if it were Ronald Reagan being shot, the space shuttle Challenger exploding, John Kennedy's Lincoln making its sad way along the Dallas streets in 1963, or that heat-seeking missile in Gulf War I, zigging through the dark and hitting its target over and over again, sometimes in slow motion.

It was hard to miss the glee with which the old media ran that clip.[3]

In the days and weeks that followed the end of the Dean campaign, the judgments against the governor and his army of followers were harsh.

His brief burst of momentum had been a fluke. A blip. Most of all, it had been just another Internet fad, a dot-com crash—long on capital, short on substance.

This is simply wrong. It was, in fact, a dot-com miracle.

In fact, it was a stunning victory that will resonate long after the election of 2004 is forgotten.

In fact, it was the opening salvo in a revolution, the sound of hundreds of thousands of Americans turning off their televisions and embracing the only form of technology that has allowed them to be involved again, to

---

[3] Of course, the Internet played a big part in spreading the "I Have a Scream" speech too, proving in the most perverse way just how powerful the medium had become.

gain control of a process that alienated them decades ago. In the coming weeks and months and years, these hundreds of thousands will be followed by millions, and this revolution will not be satisfied with overthrowing a corrupt and unresponsive political system. It won't stop at remaking politics. And it won't pay attention to national borders.

In fact, if every business and civic leader in every sector of the economy and in every segment of society doesn't think that in the next decade they're in for Howard Dean-style surprises from the people they've been treating with total condescension, they haven't been paying attention. Every business that spends $20 million on television advertising and just $20,000 to post a static web site that is updated once a month had better watch their backs. Every institution that doesn't understand that the technology is finally here to allow people to reject what they're being given and *demand what they want* had better start paying attention.

The revolution comes for you next.

When the Dean campaign ended and I sat down to write this book, several people asked if it would be a standard campaign memoir, a tell-all with all the juicy behind-the-scenes details about what went right and went wrong during Howard Dean's dramatic rise and sudden fall.

These people still don't get it.

The truth of this campaign, the "tell all," the juicy behind-the-scenes details are these: a woman who sold her bike for democracy and inspired hundreds, maybe thousands of people to do the same; a man who raised $400,000 in one week by himself by doing nothing more than sending out an e-mail; an eighty-nine-year-old man who said that he thought he was done living until the Dean campaign re-engaged his life with meaning and civic purpose.

Yes, this book is the story of a long-shot presidential campaign. But it's far more than that.

For me, it's the story of a person who spends his life reconciling two vastly different worlds—politics and technology—and wakes up one morning to find himself standing at the place where they're about to converge, to crash together and begin reversing fifty years of political cynicism in one glorious explosion of civic re-engagement.

It's the story of dozens of committed people who waged a political campaign unlike any in history. It's about the things that we did right, the mistakes we made, and the lessons *we* learned that can be applied to every election, every product, every issue in America. It's about the man we rallied behind, a politician who had the courage to stand up and question the country's path when all the others seemed to want nothing more than to hide.

But most of all it's the story of people standing up and making themselves heard. It's the story of how to engage those Americans in a real dialogue, how to reach them where they live, how to stop *selling* to them and start *listening* to them, how to make better use of the most revolutionary idea to come along since the first man learned to light a fire.

No, I'm not talking about the Internet. Or computers. Or telecommunications.

I'm talking about democracy.

PART I

GET ON THE PLANE

1

THE BEGINNING
Planes, Politics, and Pez Dispensers

I WAS BORN right when everything started going to hell.

It was 1956, full dawn of the television age, when the number of households with televisions topped 75 percent and when, not coincidentally, American political and civic involvement was beginning its long downward spiral. In my lifetime, television has become the dominant force in American life, affecting every part of our culture. At the same time, it began to erode some of the political and social underpinnings of the greatest civilization in history. If the Greeks were a people destroyed by hubris, the Aztecs by brutality, and the Romans by arrogance, Americans at the turn of the twenty-first century were a culture in danger of being ruined by *Must See TV*. Television's impact has been so overwhelming, so insidious, that it is impossible for some people to imagine a world not dominated by it, to believe that something new could rise up and break TV's fifty-year spell of cynicism and powerlessness.

But I have seen it.

And so have you.

We saw it when an army of nineteen-year-olds used Napster to bring the recording industry to its knees. We're seeing it in corporate America, where small investors are beginning to band together on web sites and blogs to demand accountability from the companies they own shares in. We're seeing it with TiVo and *American Idol* and a flood of reality programming, as television desperately tries to remake itself in the image of the Internet. We saw it in China, where citizens used the Internet to get their government to confront the AIDS epidemic and in the Philippines, where demonstrations organized by text messaging drove out the president of the country.

And we saw it most recently, in the United States, in Howard Dean's insurgent bid for the presidency.

For twenty years, people have been calling this era of computers, the Internet, and telecommunications the "information age." But that's not what it is. What we're really in now is *the empowerment age*. If information is power, then this new technology—which is the first to evenly distribute information—is really distributing power.

This power is shifting from institutions that have always been run top down, hording information at the top, telling us how to run our lives, to a new paradigm of power that is democratically distributed and shared by all of us.

I believe that what we do with that power will determine the course of this country. I believe that the Internet is the last hope for democracy. I believe that Americans will use it in the next decade to bring about a total transformation of politics, business, education, and entertainment.

Personally, I can't wait. But then, I've always been the kind of person who thrives on change, the kind of person who runs headlong into things I don't understand . . . even those things that scare me to death.

Ever since I was a kid, I've had this recurring nightmare. In the dream, my friends and family are frantically telling me that I have to get on an airplane. They won't tell me why, but they are adamant that I do it: "You gotta get on that plane. It's important. You have to get on that plane."

I was seven or eight when I first began having this dream. It was so vivid, so real—and it was the same every time: *"You gotta get on that plane."*

In the nightmare, the same thing always happens: the plane sputters and rocks and eventually goes into a dive, hurtles toward the ground, and blows up in a big fiery explosion. As befits someone who wakes up every other morning having just died in a plane crash, I grew up terrified of airplanes and of flying.

I think that most people, when confronted with this kind of phobia, would find a way to avoid flying at all costs.

Not me. I became obsessed with flying. Maybe as a way to conquer my fear, I learned everything I could about planes. I learned why they fly, how to take them apart, and how to put them back together again. I spent hours hunched over models. I read every book and magazine about planes that I could get my hands on. I went to the airport to watch planes take off and land. When I was old enough, I got jobs at airports working around planes. And, eventually, I chose a career that meant I'd be on airplanes most of my adult life.

In the end, I believe this is the only effective way to deal with an irrational fear like that—to put your mind to work on the problem, to turn the fear of flying into an understanding of it, a sense of wonder at the miracle of human flight.

I understand the fear of communication and information technologies. I know there are corporate and political leaders who, to this day, refuse to acknowledge the immense and mostly untapped potential of the Internet. I know why you belittle technology, why you call the Internet a fad, why you comfort yourself by believing that we left all that nonsense behind in a speculative stock bust. I know what motivates you to say that it costs too much, that it's going to open the world up to new problems, that if you jump too soon into this new technology, you'll fall out of the sky.

You're afraid of this force you do not understand and cannot control. You'll check your e-mails, or run a search engine, but you'd like the Internet to remain static, like television, where the shows might change, or the screen might get bigger, but for fifty years, the damn thing has generally stayed the same. It's like the person who swallows his fear long enough to get on a plane, but doesn't want to think about why it stays aloft. Always lurking behind that fear is ignorance; once you know why the plane flies . . . you'll also have to confront why sometimes it doesn't.

It's natural to fear things we don't understand. But in this era of warp-speed technological discovery that we've been in for more than fifteen years—when gadgets seem to become obsolete before you've finished reading the manual—fear of change (and its related fear of *changing too soon*) is a terminal condition.

I've got news for you politicians and business people and anyone else who has let your fear of technology keep you from understanding and embracing the Internet. This is the dominant technology—not of some distant future—but of tomorrow, of next week, of now.

And you're almost out of time.

The quantum steps are coming faster now and the rate of change is about to become blinding. The stop and start of the dot-com boom and bust will seem like nothing more than the jet engines stalling, and then rumbling to life, and anyone who doesn't admit that this technology is in the process of transforming America is going to miss the takeoff.

As I told Governor Dean in the beginning of his unlikely campaign for the presidency: In order to win, you can't be afraid of losing. Before you can fly, you have to get on the plane.

THE FIRST CAMPAIGN

I grew up in Los Angeles in the 1960s and early 1970s. My parents split up when I was three and my father lived four hundred miles to the north, in San Jose. My mom remarried and divorced again, so I lived with my two brothers and two sisters in a tiny house on the wrong end of Sunset Boulevard. We were very poor—hand-me-down poor, water-on-your-cereal poor—but my tireless mother, Peggy, found a way to make it work, supporting five kids by working two waitress jobs, at Denny's in the morning and at a nightclub until well after midnight.

It was in that house on Sunset that I first found my passion for airplanes and technology. It was also in that house, crouched in front of our black-and-white television set in 1968, when I was eleven, that I had my first vision of politics, of its power, of its ability to inspire and transform people, and also of its potential for tragedy.

I can still see Bobby Kennedy walking into that packed ballroom at the Ambassador Hotel in Los Angeles—just a few miles from my house—and making his victory speech ("Now it's on to Chicago and let's win there!"). I was cheering in front of the television as he was led out of the ballroom through his supporters, and into the kitchen, and I was stunned watching the confused reaction—people running, pointing, crying—after Sirhan Sirhan shot him.

Our neighborhood was very poor and racially mixed: equal parts black, white, and brown; one of those neighborhoods where there was so much color no one much noticed, or cared—and where the newest toy was any baseball that still had its stitches intact. When you grow up surrounded by poverty, social justice and democracy are not abstract ideas. Hunger and joblessness are not just statistics when you see people lined up for welfare and unemployment checks. I watched my own mother work stubbornly to do everything she could to stay out of the welfare line—and then watched her cry and think she'd failed when sometimes the only way to put food on the table was to relent and stand in that line. Years later, when Ronald Reagan yapped on and on about Welfare queens in their limousines, my skin crawled.

While I wasn't overtly political until I was in college, the Los Angeles I saw growing up reinforced the sensation I had watching Bobby Kennedy's assassination: the world was an unfair place that needed improvement, and every once in a while someone special stepped up to do something about it—sometimes at great sacrifice.

In high school, I ran track and occasionally found myself in class, where I worked just hard enough to remain athletically eligible. And so my grades left something to be desired (unless you desired Cs). I had a friend named Marc Cobb, a black guy who lived in Watts and ran track with me, and one day Marc mentioned that he was getting ready to take the Standardized Achievement Test to get into college.

"The what?" I asked.

So Marc dragged me down to take the test with him. I often wonder about Marc, if he has any idea that he saved my life. Surprisingly, mostly to me, I crushed that test, scored around 1500. Suddenly colleges were contacting *me*.

I chose San Jose State University, where I wanted to study aeronautics. I put myself through college, working after class at a pizza place and as an all-night desk clerk at a seedy hotel. I also drove a delivery truck for my father, who had moved his florist shop to San Jose. He was a tough old Sicilian who considered college a waste of time and fully expected me to take over the business when I was done messing around with school.

One day during my freshman year, in 1974, I was in the cafeteria minding my own business, when I saw a guy with hair down to his ass, walking from table to table. At each table, he'd stop and say something, and the people would shake their heads no. Finally, he got to my table. He explained that his name was Dennis Driver and that he was running for student body president as part of a group called the *Alliance for a New Democracy*. But to run for president, he had to have a whole slate of candidates on his ticket and he needed one more name for the student council.

"Can I put your name down?" he asked.

When I looked skeptical, he said, "Look, don't worry about it, man. You won't win or anything. I just need a name for this slot."

I let him put my name down and it turned out Dennis was right. The Alliance for a New Democracy was a lark, the fringest of fringe parties. In fact, of the twenty or so names he had on his clipboard, there would be only one winner.

Me.

I had made the first political blunder of my career: underestimating the value of a candidate with the last name *Trippi* on a college campus in the year 1974. The only candidates that I could possibly have lost to would have to be named Jimmy Acid or Tom Mescaline.

So I was on the student council. The next year, I was the student body vice president. Politics was my new airplane. I loved it. I became a shaggy-haired activist, leading the fight for campus parking, starting an independent newspaper, and reviving an edgy, defunct literary magazine.

One day, I was reading the daily San Jose newspaper, the *Mercury News,* when I noticed a front-page story about a city council race. The gist of the story was that this longtime councilman, Joe Colla, was essentially assured of retaining his seat because the filing date had passed and no serious challengers had surfaced. The least threatening of his opponents was an African American woman named Iola Williams. The story implied, at least to me, that Iola couldn't possibly win in a city like San Jose, which was only 3 percent black.

I can't tell you how much that story pissed me off. *She can't win because she's black?* This was 1975 in San Jose, California. Not 1955 in Savannah, Georgia. I was outraged. So I got on the phone, called Iola and asked if I could walk a precinct for her. Then I waded out among the tidy ranchers and split-levels in San Jose and knocked on doors, asking people to support Iola Williams. A couple of days later I did it again. There were only two hundred and forty precincts in all of San Jose, and I realized that if I could just get two hundred students to go with me, we could walk the entire city of San Jose in a day. So that's what we did.

Iola didn't win, but her surprising showing kept Joe Colla from reaching 50 percent, and forced a runoff with the guy in second place, Jerry Estruth. Iola called Jerry and offered her endorsement, but asked as a condition that he hire me to work on his campaign. So I did the same thing, got the college students to walk the city, this time for Estruth, who beat Colla. And when another city councilman was forced to resign after a scandal, Iola Williams was appointed to his position and, a few years later, would go on to be the first African American elected to the San Jose City Council.

Around the same time, the president of San Jose State, John Bunzel, was giving a series of speeches about the Alan Bakke reverse-discrimination suit, which was before the U.S. Supreme Court. Bunzel argued that Bakke had been wronged by the system, that affirmative action was a horrible thing.

Before I knew it, I was on the front page of the *Mercury-News*, leading a student movement calling for Bunzel's resignation. We took our case to the city council, which called for Bunzel's resignation. We petitioned the governor.

Finally, in 1978, John Bunzel resigned from San Jose State University.[1]

Not long after that, I also resigned from San Jose State University. Still a few credits shy of graduating, I was twenty-two years old and had already helped win two seats on the City Council, and was a key organizer in a drive that ousted a university president. I knew what I wanted to do with my life: work fulltime in politics.

About that time I was contacted by a bright, well-spoken, and preternaturally mature guy named David Bender (one of those people you're surprised to find out is only a year older than you) to organize San Jose students in the Draft Kennedy movement, to get Senator Edward Kennedy into the 1980 presidential campaign. When Kennedy finally announced his candidacy in 1979, I called David, who was the California campus organizer: "David, you gotta get me into Iowa. I want to do advance for Senator Kennedy. This is all I want to do in the world. I want to get people out to vote, make sure they turn out."

---

[1] He would resurface in 1980, when Reagan was elected president and appointed Bunzel as the chairman of his civil rights commission—an Orwellian appointment if ever there was one: a civil rights commission chairman who didn't seem to believe in civil rights.

"Joe," David said, "that's not *advance*. Advance is when you go out ahead. What you're talking about is *field*."

At that point, I only knew two guys in politics, both of whom had recruited me for the Kennedy campaign, David, and an outgoing young advance guy named Jon Haber. If Jon was advance, that's what I thought I wanted to be. "Look, I don't give a shit what it's called," I said. "I just want to do it."

A few days later, the call came. Get my ass to Des Moines, Iowa. I was ecstatic.

Months earlier, in my role as a campus agitator, I'd led protests against Bank of America for supporting apartheid by doing business in South Africa. Unfortunately, a row of businesses near the campus in San Jose got caught in the crossfire—businesses whose owners had loans with Bank of America and who wouldn't have known South Africa from South Dakota, and couldn't understand what the ruckus was about.

One of those businesses was my father's flower shop.

When my dad asked me to stop, I told him that I was fighting for what I believed in. He looked at me and said that people at the bank were asking him, "Can't you control our own son?"

It hurt, but I looked at him and shot back, "Tell 'em no, you can't."

So he was already furious with me when he found out I was dropping out of school to go work on a campaign, to be—in his words—"a bum" and "a hack."

In late 1979, I packed my things and left San Jose, out of the florist business forever. It would be five years before my father and I would speak again.

DAWN OF THE GEEKS

I wanted to change the whole damn world.

Nothing short of that was worth getting out of bed in the morning. Whether it was taking on my own university president over affirmative action or defeating apartheid in South Africa, the only direction to move was forward: Do it better, fix problems, advance. That was the attraction of politics. Bobby Kennedy's message was one of progressive democracy— not governing for the sake of governing, but bringing everyone forward, lifting the poor and oppressed, raising the quality of life for all Americans.

If politics isn't a force for change, if it isn't committed to putting poverty and crime and racism and war behind us, then what good is it? If politics can't improve the world, then why the hell bother? I suppose that was the appeal of the Democratic Party—it moved forward while the other side was conservative and reactionary, intent on keeping the status quo—or worse—rolling back the clock.

I was frustrated by the pace of change in America, and outraged by resistance to it. I desperately wanted to see what tomorrow looked like. If there was some method of getting there faster, some tool to involve more people, then I wanted it. Years later, when we first began using the Internet to propel Howard Dean's presidential candidacy, people were shocked that technology could play such an important role in the regressive, always-done-it-this-way world of politics. What did a bunch of geeks hunched over keyboards have to do with running the country?

For me, politics and high tech have always sprung from the same well—the balls-out desire for progress, the idea that the greatest force for political and social change in America has always been the ingenuity and creativity of its people, and that if you bring enough of them together and give them the tools—a petition, a candidate, a computer—they could transform the world.

From the beginning, I was a child of these two fathers—politics and technology. And if San Jose State in the mid-1970s was where I became enthralled with the world of political organizing, it's also where I became a full-fledged technology addict.

Even after I began to lean away from aeronautics as a career (I couldn't really see spending the rest of my life in some room figuring lift drag coefficients), I continued to loiter around the San Jose State physics department, randomly signing up for the most speculative science classes available . . . like Laser Holography. I had no idea that the primitive work I was doing—shooting a red-beam laser off the face of a quarter to capture a shiny three-dimensional image of George Washington in a plate of glass—would be the same process thirty years later used in the holograms that appear on credit cards to foil counterfeiters.

As it turned out, San Jose was one of the birthplaces of the coming computer age, and so everywhere I turned someone was doing something fascinating that I had to find out about. I was especially drawn to the fledgling computer science labs, with their huge proto-computers—room-size reel-to-reel behemoths that ran on IBM index cards and didn't have the processing power that your cell phone has today.

The first generation of geeks was everywhere, walking around Bay Area campuses, staring off into space, mumbling computer code, already beginning to plot their binary takeover, and I moved as easily in their world as I did in the campaign headquarters of Iola Williams.

Okay. Full disclosure: I was one of them. Beneath my bushy hair and denim jacket, I was just another gadget-obsessed geek, hunched over a computer screen, repairing pinball machines for extra money. The minute some new device came out that promised to help change the world, I had to have it, even if I was the only one.

It's a classic problem with technology, the rate of penetration for the earliest users. For instance, when the telephone was developed in 1876, it was clearly a groundbreaking invention, but it would be nearly *thirty years* before 10 percent of Americans had a phone in their house. For decades, neither Alexander Graham Bell nor the phone company had any idea how powerful their new technology would become because they hadn't figured out how to get it into the hands of users yet.

Of course, if I'd been around in 1880, I would have been one of those four people with a phone, sitting in my farmhouse staring at the thing I'd just traded my tractor for, wondering when the other three people were going to call me. I was the guy who bought the $2,000 calculator (with half the functions of today's five dollar digital wristwatch). I raced down to Radio Shack and bought the first TRS-80 portable computer (affection-ately called the Trash-80 by its users, reporters for big city newspapers and those sleep-deprived guys who hung around Radio Shack all day) and snugged the telephone into the little cuplet to transfer information . . . over *the phone lines.* I was even the proud owner of an Apple Newton, the much-beloved, ultimately failed, brick-sized precursor to all the Palm Pilots and PDAs. The technology graveyard is filled with gadgets that arrived two or three years too early and I owned almost every one of them.

I also remember, in the 1970s at San Jose State, staring over the shoul-ders of my friends into a primitive computer monitor and watching a bunch of numbers scroll across the screen, computer code that was being shared with a guy somewhere else in the country staring at his computer, over this fledgling network called the ARPAnet, an old defense department com-puter network and the precursor of what would later become the Internet.

From the moment it was first drawn up as a way to increase computing power by connecting researchers' computers, the Internet was a singularly democratic and open network. The initial operating system for this network—before Bill Gates and Microsoft had the darkly brilliant idea of licensing and selling this stuff—was open and its source code was available to other programmers, who were invited to mess around with it, to see if they could improve it and pass the new-and-improved version on to the next person to tinker with and improve some more. As the technology guru Howard Rheingold wrote in his book *Smart Mobs,* "The Internet was deliberately designed by hackers to be an innovation commons, a laboratory for collaboratively creating better technologies."

This ground-up architecture is what sparked the dynamic growth in computers and, eventually, the Internet—this open invitation to improve, to innovate, to make it better. Anyone could contribute to this system, and anyone did: government, corporate and university researchers and soon—fifteen-year-old kids in their basements.

All over San Jose, all over the country, people were envisioning not the world that was—a bunch of fuzzy numbers flashing over black and white screens—but the world that *could be:* a network, a web of computer users sharing information: life stories and computing tips, recipes for cookies and petitions to abolish apartheid.

It was a thrilling, intoxicating time, and I sometimes wonder what would have happened if, instead of deciding to indulge my political half, I'd stayed in Silicon Valley with my computer peers, like one friend from college who soon found himself working with a handful of other people to find other computer users who might be able to help his friend's wife buy old Pez dispensers.

The little community they built, eBay, now has about forty-five million registered users. This year they will buy and sell $52 billion worth of Pez dispensers, baseball cards, sewing patterns, and whatever else strikes their fancy. But eBay is more than a multibillion dollar business; it is a community, a city of people broken into neighborhoods by hobby and interest and commerce, a community of people who, if they banded together, could rise up, and in a single day change the world all by themselves.

2

THE FIRST CAMPAIGN

Jimmy Hogan, Irv Gadient, and the PDP-11

THERE'S ONE THING that most people don't understand about a presidential campaign:

Everything.

Working day-to-day on a presidential campaign is unlike any other job on the planet. It's a thankless job, an outrageously difficult job, the most emotionally draining, physically taxing, stress-creating job you can imagine, and when it's done, it almost always ends in total, abject failure. Since 1968, sixty-three serious candidates have run for president on the Democratic ticket. Two have won. That's two in thirty-five years. The other sixty-one have gone home broke, beaten, and exhausted. This is hard enough on the candidate, but for the staff—working twenty-hour days with no days off, desperately trying to do the impossible, and dying with every misstep—it's the psychological equivalent of self-mutilation.

There's no glory in working on a presidential campaign. Whatever glory there might be goes to the candidate. There's no money in it. No prestige. No perks. You sleep on motel beds, or on couches or in cars, if you get any sleep at all. Dinner is a slice of pizza or a race through a drive-through window.

Working day-to-day on a presidential campaign is undoubtedly the stupidest human endeavor ever conceived.

And I believe everyone should do it once.

In November 1979, when I left San Jose, California, and drove my gold Ford Pinto 1,800 miles to Des Moines, Iowa, I was eager to do whatever I could to get Ted Kennedy elected president. This was not going to be easy. Kennedy had what we now call "character issues," and was the nation's preeminent liberal at a time when America was taking a sharp right turn. If all that wasn't enough, he was also running against an incumbent president. Even though President Jimmy Carter's popularity had fallen, it was still unheard of to defeat a sitting president from your own party, and to attempt it was considered political blasphemy.

But this was a Kennedy, the brother of the man who'd inspired me to get into politics, heir to the throne of liberal Democratic politics in America. I was thrilled and honored. I did what every foolish kid does when he gets to work on his first presidential campaign: I did somersaults down the halls.

My first assignment in Des Moines was to go to the airport and pick up a couple of other field organizers who were flying in. One of them was a bright, young guy named Tony Pappas, who until a few days earlier had worked in the Carter White House. People can work their whole lives and not get where Tony had gotten, but he'd just quit this prestigious, decent-paying position to work long hours in the field for Ted Kennedy for fifteen bucks a day, hoping that by some miracle he could get Kennedy, and maybe himself, to the White House . . . the very place Tony Pappas had just quit!

Right away I felt at home in this world of irrational commitment, with this collection of cynical idealists.

I knew all about the gonzo reputations of campaign organizers and career pols. In college, I'd devoured Teddy White's *Making of the President* books and, especially, Hunter S. Thompson's *Fear and Loathing on the Campaign Trail*. I was enthralled with Thompson's description of a field-organizing legend named Gene Pokorny, a self-described "grassroots gunslinger" for presidential candidate George McGovern in 1972.

The guys I met on the Kennedy campaign were all from that mold— whip smart, caustically funny, tireless, fearless, and larger than life, just one shot of adrenaline from going over the edge. I felt like I'd found my long-lost tribe, the only people who thought and acted the way I did. We were called the Corn-Stalkers. Some of the guys were already big names in political organizing—like Carl Wagner, a contemporary of Gene Pokorny's;

John Sasso, who would go on to mastermind the Michael Dukakis campaign in 1988; Paul Tully, a gentle bear of a man and a former football star at Yale, who had worked for Eugene McCarthy, George McGovern, and Morris Udall; and Mike Ford, an Xavier graduate who cut his teeth on campaigns for Udall, Jerry Springer,[1] and Birch Bayh, and who was just as big, just as eclectically brilliant, and just as totally nuts as Tully.

They called each other "Brother Ford!" and "Brother Tully!" or "Bro" for short, and there was nothing these guys wouldn't do for you, nothing they wouldn't do for each other, nothing they wouldn't do for a vote, and nothing they wouldn't do for a meal. I don't know how many times I saw Tully and Ford, at two in the morning, rumble in carrying eight large pizzas, ready to break the arms of anyone who even *thought* about asking for a slice.

Every campaign is split between the people who come in with the candidate—usually trusted aides from his governmental staff—and the pros, campaign freelancers who drift off to the next campaign once the candidate wins (or loses). In general, the members of these two groups automatically assume the other group is made up of idiots. Even in the best campaigns, you view the other side with a healthy disdain: the coarse, unpredictable, antagonistic pols vs. the ambitious, overly cautious desk jockeys. I could see right away where I belonged—with the Fords and Tullys of the world.

Later, when the book *Quest for the Presidency 1984* included my name in the same sentence as those guys—". . . blooded 1980 campaign veterans like Paul Tully, Mike Ford and Joe Trippi . . ."—it was the most shocking and humbling thing I'd ever read. These guys were legends. I was . . . well . . . I was me. Seeing my name after those guys' was like a career utility infielder coming across his own name mistakenly listed at the end of a great baseball lineup: Gehrig, Ruth . . . Trippi?

Back in 1979, about the only thing I had in common with Ford and Tully was that we all happened to be in Iowa. And unlike those guys, I didn't know my ass from Des Moines.

Right away, we were assigned counties. Mike Ford was given Waterloo, Steve Murphy, who would emerge as another lifelong friend and a star in campaign organizing, Council Bluffs. I was given small, out-of-the-way,

---

[1] Yes, that Jerry Springer. Before he was a talk show host breaking up fights between adulterous transgender couples, Springer was the mayor of Cincinnati.

conservative Jones County. Then we were each given a list of the known Kennedy supporters in each of these counties.

I watched the county organizers get their names in boxes: Three boxes of names for this county. Two boxes for that county. Then came little Jones County.

My boss, Jack Corrigan, handed me an index card.

I turned it over in my hand. "What's this? Where's my box?"

"That's it," he said.

I looked down at the card. There was one name on it: Irv Gadient.

There were 20,401 people in Jones County, Iowa, and the only known supporter of Senator Edward Kennedy's presidential candidacy was a guy named Irv Gadient? That day, I drove up to Irv's house and put my stuff in his spare bedroom and sat there, staring out the window at miles and miles of flat, snow-covered farmland, while the other 20,400 people waited to see what Joe Trippi was going to do. I had never seen snow. And I didn't have the first clue about where to start organizing a county.

As I told a group of Dean field workers in Iowa in 2003, I was petrified. I stood around for three days, staring at my shoes, sure that at any moment I was going to be discovered as a fraud.

And then I got on the plane. I used my fear to push myself into action, to conquer the thing I feared by taking it apart, by figuring out how it worked, by teaching myself on the fly how to organize a county in the Iowa Caucuses. Or to crash in the process.

I started in the small town of Olin, population 700, maybe a third Democrat, almost all of those Carter supporters. But I quickly found out there was a war going on in Olin. Cavey's Café on Jackson Street had recently begun carrying *Playboy* magazine and the town was split down the middle between those who thought it was about time *Playboy* came to Olin, and those who thought it was the first sign of the Apocalypse.

Given Ted Kennedy's negatives as a candidate, I doubt my superiors would've approved of my asking Dave Cavey to be the Chair of the Olin Kennedy campaign based on the fact that his café carried *Playboy*. But having now doubled my voter base to two, I began to work tirelessly, putting up signs, knocking on doors, wheeling and dealing and cajoling and, when necessary, driving my Pinto into a ditch and striking up a conversation with the farmer who came by in his tractor to help me out.

The Democratic chairman of Jones County was a crusty Carter supporter named Jimmy Hogan, who lived across the highway from Irv Gadient. Back then, in these rural towns, the caucuses were held in people's homes and so, if a Carter supporter like Jimmy Hogan hosted the caucus, then out of a sense of Midwestern propriety—which trumps both political parties in Iowa—Kennedy supporters wouldn't think of going to a fella's house to disagree with him. The Kennedy voters would stay home and Carter would walk away with that caucus.

In a place like Jones County, Iowa, you get Jimmy Hogan and you were halfway to delivering the whole damn county.

This is something not everyone realizes about our political system. Not all voters are created equal. Some people carry more influence. In his book on consumer epidemics *The Tipping Point,* Malcolm Gladwell writes, ". . . the success of any kind of social epidemic is heavily dependent on the involvement of people with a particular and rare set of social gifts." Gladwell calls these influential people, people like Jimmy Hogan, "connectors."

In the Dean campaign, we called them *bloggers.*

But in 1980, I wasn't going to reach Jimmy Hogan with an e-mail or by posting on an electronic bulletin board. So I had no choice but to drive up his long dirt driveway and ask if he could somehow let his neighbors know that, even if they wanted to vote for Kennedy, they were still welcome in his house the night of the caucus.

Jimmy was a craggy old farmer, too busy to listen to my arguments. He pointed out that he had plenty of dairy cows that needed his attention a lot more than I did. For six weeks, I bugged him. Finally, he told me that he milked his cows every morning at 5 A.M. If I wanted to talk to him, I could show up then, but I'd better be ready to milk.

So I began showing up on Jimmy Hogan's farm at five in the morning, milking cows and pleading for him to open his house up to Kennedy supporters.

Finally, with ten days to go before the caucus, he agreed.

Most people don't really know how a caucus works, especially if they have the same experience with voting that I had in California: step into a voting booth, pull a curtain, punch a hole, and leave. All over the country, people do the same thing—casting their votes one at a time.

A caucus is a very different affair. It is grassroots politics at its most basic level, a bunch of neighbors getting together every two years to talk about their common interest—in this case politics—and, during a presidential campaign, to choose the candidate they prefer from the two major parties. They shoot the shit about farming and their families and the local basketball team, and then get around to talking about the issues that are important locally and the issues they'd like to see the party pursue. And then they divide into sides, count the numbers for each candidate, and assign delegates based on that number.

A caucus reveals something profound about our political system, the importance of the social and economic fabric that has always existed beneath our political systems—the very reason I had to ask Jimmy to welcome Kennedy supporters into his home. It is a system that reflects the way people live, the way they relate to their family, to their jobs, to their interests, and to their communities. The Iowa Caucuses are a participatory, social, bottom-up civic engagement—everything that politics has lost through its black-tar addiction to television advertising, and everything that the Internet offers again: a chance for people to not just vote, but to become involved again, to write the agenda, to contribute to the organization, to affect more than just the numbers. A business owner trying to understand how to reach a community of people on the Internet could do worse than to attend an Iowa caucus.

This is one of the key reasons I think that the Dean campaign was able to make good on years of commercial theories about the Internet in an area that few people expected. Political organizers like myself had years of experience with the very issues that create so much potential for growth on the Internet, issues that also make it difficult for the usual American corporate giants to control.

The night of my first caucus, in January 1980, I went to Jimmy Hogan's house at 6:30 P.M., and watched his neighbors pull up in their pickup trucks and American sedans.

Jimmy's living room was packed. When it came time to call for the vote, the two sides began moving to opposite corners of the room, an old flickering chandelier demarking the line between them. I watched the people move to their corners, doing the math in my head. It was close; a few votes either way could turn it. And just as they were about to start counting, this seventeen-year-old girl—Jimmy's daughter, or maybe his niece, I can't remember—began a slow determined walk across the room.

In Iowa, if you're going to turn eighteen by the time of the general election, you can vote in the caucus. I watched, stunned, as this teenaged girl marched across the room, leaving the Carter supporters to go over to the Kennedy side. As she reached the chandelier, I started thinking, my God, if we could win in Jimmy's house—

"Young lady," Jimmy's voice boomed across the room. "You take one more step and you're grounded for the rest of your life." I don't know if she had any ballet training, but at that moment she performed a perfect pirouette under the chandelier, turned and walked right back over to the Carter side of the room.

Of course, all over Iowa, things like that happen. You show up to the caucus meeting, see your shop foreman on the Carter side and think twice about declaring for Kennedy. You pull up, see your ex-wife's car, and just keep driving. It's a nonscientific, arbitrary, messy, heartbreaking system. And it's absolutely vital to our democracy.

That night, as the numbers started coming in, I was devastated. I had lost. I had let the senator down. When it was done, we'd dropped Jones County by three hundred votes, which translated to four percentage points, 52 percent to 48 percent. Here I'd been working day and night, cutting deals with Dave Cavey, milking cows with Jimmy Hogan, plotting and planning with Irv Gadient, driving my car into ditches all over eastern Iowa, knocking on a good share of the county's 6,931 front doors. And for what? To lose?

I called Jack Corrigan to dutifully report my failure. I was in agony. "I'm sorry, Jack. I let you down."

"Gimme the numbers," he said gruffly. "How bad is it."

I told him.

He was quiet for a moment. Then he asked me to repeat the numbers. I did.

"Joe, you only lost by four. Do you realize we're getting creamed all over the state? We're getting beaten by thirty points. You only lost by four? In Jones County? Hell. You did great, bro." In Kennedy organizing circles, there was no greater way to honor someone than to call them *bro* or *sister,* and now I had joined the ranks of Tully and Ford and Corrigan—somehow in losing Jones County, I had become a brother. And I'd learned something important about politics Kennedy-style: The cause was everything. Win or lose, spend yourself completely, leave nothing on the table, not even your health. Losing would be painful, but not as painful as knowing there was something else you could've done.

In Jones County, there was nothing more I could do. The final numbers for the state were as grim as Jack promised. Jimmy Carter beat Ted Kennedy by 28 points that night in Iowa. He won 98 of the 99 counties. Only three counties in Iowa did better for Kennedy than Jones County. The fraud police would have to come for me some other day.

I had found a home in the insular, adrenaline-driven world of political organizing. After Iowa, in quick succession, I worked my ass off on the ground in Maine, New Hampshire, Illinois, Arizona, Texas, and Michigan. Of course, Carter swept past us to victory in the Democratic primaries, only to lose to Ronald Reagan in a landslide in the general election. When it was over, I was empty. I had never been so tired in my life. I was certain about two things: It was the greatest thing I'd ever been involved in. And I would never do it again.

DROPPING PEBBLES IN THE WATER

The 1980 presidential race over, I began a cycle I'd continue for the next eight years: doing organizational and field work for other races—members of congress, mayors, governors, city attorneys—and vowing that I would never, ever, as God is my witness, take another presidential. And I meant it.

In 1981, I went to work for Tom Bradley, the mayor of Los Angeles, who had decided to try and become the first African American governor of California.

Political organizing is all about finding those people who think like you do and drawing them into your organization any way that gets them involved—everything from canvassing to donating money to simply voting for your candidate—while at the same time trying to get your message out to the people who haven't decided yet.

In that way, it's not so different from Ford Motor Company, or Burger King, both of which need to keep serving those people who already *like* Ford F-150s or Whoppers. But they also need to draw in the guy who doesn't know what kind of truck he wants, or the person who is hungry but doesn't know what sounds good, or even the person who didn't know he was hungry or never imagined he wanted a truck.

The difference is that since television began revolutionizing and dominating advertising and the rest of our culture, Ford and Burger King have simply thrown their ads on the air and waited to see who shows up. This is top-down, one-to-many advertising and for the last half of the twentieth century, it was the only way to do business.

But for 150 years in America, politics worked in the other direction, on the ground, from the bottom up. Even when I started, in 1980, a good portion of the job was still done at the grassroots level. This was the part of politics that I loved and the part that I was best at—the field work, meeting with people, getting them involved, navigating the real-world communities, neighborhoods, labor unions, and other places where people intersect with their political lives.

As a field organizer, you're always looking for tools to help you identify your voters. On the Kennedy campaign, I had set up a lemonade stand with a sign offering free lemonade to Kennedy voters, with the idea that rather than make fifty thousand phone calls, maybe we could get them to come to us.

On the Tom Bradley campaign in 1981, still living in the two worlds I'd inhabited since college, I made what I thought was another innovative suggestion.

We needed a computer.

They just stared at me. It was like I'd suggested we buy a submarine.

When I couldn't get anyone at the Bradley campaign to agree to spend $17,000 for a computer, I went into debt to buy one myself and leased it to the campaign for $500 a month. I can't say for sure, but it had to be one of the first computers ever used in-house by a political campaign.

There was no such thing as "a personal computer" then. This one was a 16-bit DEC PDP-11, top of the line, as technologically advanced as you could find in 1981. This meant it looked like a washing machine, only bigger, with a plastic, Star Trek-looking one-piece monitor/keyboard on top. It had a whopping 1.5 megabytes of memory—the equivalent of two or three great big floppy disks, piles of which were stacked like cordwood around the Bradley offices.

We used the computer to organize the names and addresses of supporters and potential Bradley voters. Even with the memory limitations of the PDP-11 system (one floppy might only hold half of the people whose name started with A) it turned out to be the perfect tool for storing and sorting names. We used it to send out direct mailers to certain precincts, including some that were overwhelmingly African American and overwhelmingly poor. It was a tremendously successful fund-raising tool, and yet the results were bittersweet, too. I still remember some of the notes that came with those $5 and $10 money orders, from people who had been overlooked by politicians before. "I believe in you, Mayor Bradley." "You give me hope. I'm not gonna eat today so I can send you this money." We had serious ethical discussions about whether we should take money from people who really couldn't afford to support a candidate. But Bradley was their candidate. We took the money.

Election night with Tom Bradley was unlike anything I've ever seen, before or since. And it was the first time I had that curse of seeing something no one else could see—even if it was only a few hours before they saw it.

In what probably was another computer first for a campaign, we had moved the PDP-11 to the Presidential Suite of the Biltmore Hotel and rigged a direct line into a box on the roof of the hotel. This connected the PDP-11 with the California Secretary of State's computer, which was tabulating results as they came in from the state's precincts. Here it was 1982 and I was sitting in a hotel room in Los Angeles, watching numbers fly across my screen—the results of an election in real time. As the polls closed, all the exit numbers had Tom Bradley winning. I was in the mayor's suite, babysitting the PDP-11 and monitoring the results when I turned on the TV and saw Nelson Rising—the brilliant chairman of the Bradley campaign—downstairs at the Bradley victory party in the Biltmore's ballroom. The television stations had all cut live to the scene, where the crowd was going crazy. A reporter asked Nelson, "Mr. Rising, how does it feel to be the next chief of staff to the governor of California?" In the room, I was so happy I cried. My God, we had done it. Against the longest odds, we had elected the first African American governor in U.S. history. And then I happened to look over at the PDP-11, the numbers dancing across its screen, counties flashing as new results hit the state's computer in Sacramento, a blur and jumble of data spilling out so fast you needed a printout to digest it. Something was wrong. It screamed out at me in those scrolling numbers—2,000 votes here, 534 votes there.

I turned to the staffer next to me. "Go pull Nelson Rising off the floor of the Ballroom—get him up here now!"

One by one we pulled key staffers aside, but many didn't want to believe my projection—hell, I didn't want to believe it either. But somewhere in that mess of digits, there was a 100,000 vote anomaly.

Tom Bradley had lost.

In the end, George Deukmejian won the governorship of California by fewer than 53,000 votes. And the greatest victory speech I have ever heard was never given.

Tom Bradley had decided if he won that night he was going to use television differently than any politician before or since. He was going to walk out to the podium of the Grand Ballroom of the Biltmore Hotel and look out at his glitzy Hollywood donors and the wealthy supporters who had funded his campaign and he was going to ask them for some patience—there was someone more important that he needed to talk to. Then he was going to look straight at the press riser and dead-ahead into all of those television cameras and ask the viewing audience on all three networks for patience—there was someone he needed to talk to even more important than them. "Please, I know it's late, but please wake up your children, I have something I want to say to them. Something important." And after a moment, he was going to read from a simple speech he had drafted himself, for California's children. "Don't let anyone tell you that you can't make it in this country. Don't let anyone tell you that you can't make a difference in this world. I am the son of a poor family of Texas sharecroppers. I was taught to play by the rules, to work hard, as hard as I could, to never ask for favors . . . and always to dream. I stand here tonight as living proof that in California, in America, dreams can come true. That someone born with no chance could stand here tonight as the elected governor of the largest state in our nation means that anything is possible—that you can grow up to be whatever it is you dream of being, and from this day forward your parents and I will work together to make sure you have the same opportunity to succeed in life that I had."

We fell short of delivering that speech by one-and-a-half votes per precinct. One-and-a-half. I watched a stoic, stately Tom Bradley slide that version of his speech back in his breast pocket, raise his head high, and deliver a completely different speech, one of concession. It would not be the last time I saw a man who represented something bigger than all of us lose. And it wouldn't be the last time my heart was broken for lack of a few votes. To this day, I wonder how much different our country would be today if that speech had been delivered. And knowing what that speech meant, what Tom Bradley would've meant as governor, it tortures me to think that there was something else I could've done to get just one-and-a-half more votes per precinct, votes that would've made all the difference in the world.

I was twenty-six years old. I had learned that no matter how noble a candidate or his campaign, the other side still kicks as hard, still runs negative ads, still spends millions to beat you. No matter how noble the cause . . . you can still lose. I came away from the Tom Bradley campaign excruciatingly aware that the more you believe in something, the more it hurts when you fail. My problem was that no matter how much politics was becoming a business, I couldn't stop myself from believing.

After Bradley, I worked a couple other campaigns and then found myself doing what I vowed I'd never do, taking the only job I have ever held in any level of government. I followed Jules Radcliff, the big, razor-sharp campaign manager for Tom Bradley, to work for the California lieutenant governor, Leo McCarthy. Jules worked as the chief of staff and I was his deputy. But months into the job, I was miserable. I couldn't get anything done. I'd come up with a good idea and have to submit it to some legislative committee, which would spring into action by tabling it. I felt like a caged animal.

The 1984 presidential race was off and running without me. Walter Mondale, who had been vice president under Jimmy Carter, was the clear Democratic front-runner to challenge Reagan, but he'd gotten an early shock from California Senator Alan Cranston, who had been hiding in the bushes, ignoring the long preparations for primary and caucus season. He had chosen instead to run a guerilla operation in Wisconsin—concentrate all his efforts to win an early straw poll (a preliminary, nonbinding vote, a cross between a caucus and a fund-raiser, in which tickets are sold to events.)

Mondale, with his huge, experienced, and professional—some might say bloodless—campaign organization, was furious. His team, finally realizing what Cranston was up to, looked up at the map and knew in an instant where the California senator would go next—Maine, the other state with a nonbinding straw vote. Mondale turned to Mike Ford and asked him if there was someone he could find who could put Maine in the Mondale column.

That day, in the fall of 1983, I was working in the lieutenant governor's office when the call came. I could tell just by hearing McCarthy's end of the conversation that it was Fritz Mondale on the phone. And I could tell they were talking about me.

"I'm sorry, Mr. Vice President. Yes, Joe is very good. But I can't let you have one of my staff guys." McCarthy had promised Cranston that he wouldn't do anything to help any of the California senator's rivals.

After he hung up, Leo thought about it for a moment, and finally, he came over. He explained that he couldn't possibly give me permission to go work for the Mondale campaign. "Of course," he said carefully, "if you were to quit and go off and do something insane on your own . . . I mean . . . what could I do? I couldn't stop you. I might even forgive you and hire you back in a few years."

I was gone the next day.

I had sworn to myself that I wouldn't work another presidential, but I couldn't stay away, and I ended up working even harder than I had in 1980. One reason was the sheer thrill of working alongside Tully and Ford again, even though we weren't exactly welcomed with open arms by the people we'd fought against in 1980 on the Carter team, many of whom had moved over to help Mondale.

The other reason I loved working on that campaign was Fritz Mondale himself, who turned out to be one of those leaders you would do anything for to help him get elected. To this day, I think that of all the presidential candidates I've ever seen, he had the best temperament, experience, and preparation for the office.

But he was in trouble before the primary season even began.

The first place we had to stop the bleeding was in Maine, where Cranston had done a number on us. The Maine straw poll conventions had a limited number of delegate tickets and he'd wrapped up more than enough to win it. We were screwed.

Looking at the jigsaw map of Maine's sixteen counties, I realized that many of the smaller towns hadn't held caucuses to elect delegates to the straw poll. But there didn't seem to be any reason why they couldn't. Ford and I had a band of young, unkempt organizers we'd trained to run on adrenaline, peanut butter and jelly sandwiches, and beer (those old enough to drink) and we raced around the state, setting up ninety-three new caucuses *in three days*. It was a fast one: Elect a bunch of delegates in towns no one knew about and hope the Cranston people didn't find out in time to put up a fight. And it worked. We elected a bunch of new delegates to the straw poll convention and caught the local state Dems and Cranston off guard. But we still didn't know if it would be enough to win.

The night of the count, Ford was on the floor of the raucous convention hall, calling in the numbers or holding up the phone so we could hear them announced right there on the floor. Up in our hotel room, the old Carter/Mondale guys were telling Fritz that it didn't look good (and to be fair, it *didn't* look good) but I sidled up to him and said not to worry. We were going to win.

The counties were listed in alphabetical order, and Ford was calling in the results of each one. *Androscoggin County . . . Aroostook County . . .* Right away, the former vice president fell way behind and the old Carter guys nodded solemnly and told him to be prepared to lose.

"Nope, you're gonna win," I repeated, drawing raised eyebrows.

With one county to go, Mondale was still behind and the old Carter guys were writing it off, saying it was over.

"Oh yeah?" I said. "Watch this sucker."

Ford called again from the floor. "Are you ready?" He held up the phone.

Each time a county read its results, it was stretched out, the way the states do at the national party conventions. We sat there, while they droned something like: *York County, the county of dogs and yellowfish and crab-flavored ice cream (or whatever the hell it was) casts 321 votes for Mondale, 5 for Cranston. Mondale is the winner!*

Fritz and I leapt out of our chairs and high-fived, but out of the corner of my eye, I could see the old Carter guys just didn't seem as happy. They'd been wrong about the outcome and the victory had come at the hands of two Kennedy guys and our young peanut butter and jelly troops. It was culture clash, pure and simple, and I was troubled by it. For me, campaigns had been about brotherhood and working for a common cause. This was the first time I'd seen two different political cultures exist in the same campaign and struggle to come to terms with each other. But because Fritz and his campaign chairman Jim Johnson trusted each other completely—and because Johnson did a masterful job of making sure both sides knew the rules (and what hell there would be to pay if lines got crossed or petty fights broke out)—the Mondale campaign held together as a strong, effective team throughout the primaries. In later campaigns, I would see what happened when that trust wasn't there, or when a strong chairman wasn't there. It almost always meant that any internal struggle would tear a campaign apart, since no single person had the power to stop it.

After "the Maine miracle," we had to scramble to get ready for the Iowa Caucuses. This time, I ran the whole state. Instead of one county, I had ninety-nine. And the pressure increased by about the same ratio.

But even running a whole state on the ground like that, it's still about those people, about the connections that you make with them. Unlike TV advertising—one-to-many—field organizing is best when it involves people in the organization, in a dialogue, back and forth, many-to-many. You work with people who go out and work with other people and soon you create a community built on respect, mutual interest, and loyalty.

I suppose that's why, a couple of times, a mysterious "high-priority stop" would make its way onto Mondale's crowded schedule while he was speaking in Dubuque or Des Moines or one of the other larger Iowa cities where the candidates spend the bulk of their time. The Mondale buses would rumble across the flat farmland and turn into a barn in the middle of nowhere.

Irv Gadient's barn.

"What the hell are we doing here?" Mondale asked when he looked at the enthusiastic, but admittedly small group of supporters in Irv's barn.

I tried bluffing: Jones County was an important cog in the overall Iowa strategy, but just then Irv came up and told the candidate how I had lived in his house for two months back when we were both for Kennedy and was like a son to him and how it was nice that I repaid him by finally bringing an actual *candidate* to Jones County, because the candidates *never* visited Jones County. Mondale shot me a look but was, as ever, gracious.

We won Iowa handily, with 49 percent of the vote, 33 points higher than our closest rival, but the victory revealed some obvious cracks in the Mondale campaign. His organization had done such a good job presenting itself as an efficient, classic Democratic machine, that the former vice president was vulnerable to the fresh ideas and youthful energy of Gary Hart, who followed a 16-point second-place Iowa finish with a stunning victory in New Hampshire.

A good political operative learns as much from his rivals as he does from his own campaign and as I watched Gary Hart, I was fascinated by his campaign. Every other candidate ran around Iowa speaking to as many large crowds as he could fit in—the way politics had always been done. But Gary Hart would pop into a town and look for that one key person. His pebble. Then he'd drop the pebble into the water. And leave. Let that energy ripple out to the other folks in town.

He described it as the politics of concentric circles—the idea of waves spreading out from a single stone thrown into the water. He'd concentrate his considerable charm and intelligence on one or two key people and genuinely connect with them. He believed that his message would spread outward from that one person better than if he diluted his message in a roomful of people.

Against the strongest Democratic candidate in decades—Mondale had every union lined up, every major Democratic constituency working for him—Hart's concentric circles spread and he stormed to victories in the Northeast and the West and was threatening to catch Mondale in New York and Pennsylvania. That's where I went next, for another burst of wall-to-wall, nonstop shoe-leather campaigning. When I arrived in Pennsylvania, we were fourteen points behind Hart. We swung the polls by twenty-eight points and beat him by fourteen. We won in New York, too, securing the nomination, only to be crushed by Reagan in November.

But I never forgot Hart's strategy. What had seemed counterintuitive at the time began to seem brilliant to me, especially for an insurgency campaign, the idea that you could spread a candidate or a cause or an issue like a virus—starting with a small, key group of people and letting them run wild for you. Unfortunately, back then there was no tool that would help you create that momentum. Not yet.

Once again, when the 1984 campaign was over, I was totally exhausted. I wanted to sleep through 1985. This time, I told myself, I really was done. Grassroots, shoe-leather organizing was thrilling, but the relentless physical exertion made it a younger man's game. After nine years of this stuff, I was ready to move to what I thought was the cerebral side—

crafting the strategy, speeches, spots, and substance of the campaign—the collective elements known as "message." Unfortunately, I was arriving at a time when the message was being completely devoured by the medium—television and its evil spawn: negative advertising, constant polling, and voter apathy. Over the next few years, as I became an expert at this cynical game, the profession that I'd once seen as a calling—politics—would begin to seem like the lowest, most disgusting business in the world.

3

THE HORROR

Thomas Jefferson, Willie Horton, and Governor Moonbeam

I REMEMBER WORKING on a congressional campaign in the mid-1980s. We were meeting with the candidate to film a commercial about his views on abortion. As we were setting up the equipment, going over the script, and getting ready to shoot the spot, the guy casually informed me that he wanted to film two spots.

I was confused. Two?

That's right. He wanted to make a pro-choice ad, then—as long as we had the crew and everything—just "turn the camera around" and make a pro-life spot. That way, he explained, he'd have one of each ready when his polling people told him which way he should go.

The crew looked at each other and shook their heads as we filmed this guy's flip-flopping ads on abortion. After it was over, I think we all wanted to go home and take a shower. For me, that day—with its vacuous, unprincipled candidate relying on polling and the slick, diminishing power of television advertising—was the epitome of what has gone wrong with politics in America.

After the 1984 presidential campaign, I went to work for Caddell, Doak & Shrum, the political media giants, writing, producing, and making television spots for a variety of local and national candidates. Pat Caddell, the mercurial, wunderkind pollster who had come out of college to lead

George McGovern's charge and had helped guide Jimmy Carter into office, left the firm not long after I arrived—in part because he was disgusted with the direction of political advertising. But over the next few years, I learned a great deal from the other two partners, David Doak and Bob Shrum. Bob and I had an especially complex relationship, respectful, often contentious and competitive, but always creative. It was as if the friction of our very different personalities sparked us to do our best work. I think back now on the string of luck I had—to be schooled in organizing and politics by giants like Wagner, Tully, Ford, and Sasso, and then to find myself immersed in the media and message side of things alongside Caddell, Doak, and Shrum, by far some of the best in the business.

My own strength on the message side lay in spending time with a candidate, listening to what he had to say, and transforming those ideas into stump speeches, slogans, and spots. A good example is the 1988 presidential campaign of Dick Gephardt.

That year, Gephardt had thrown everything he had into Iowa, hoping to repeat Jimmy Carter's 1976 strategy of riding the momentum of a surprise caucus win all the way to the White House—which had become the only workable strategy, especially for a long shot. So the Missouri congressman spent 110 days in the state, and even moved his 90-year-old mother into an apartment in Des Moines.

But his campaign was flat, his numbers mired in the single digits, and his message—which seemed to have something to do with traveling to a lot of places as a congressman—was putting voters to sleep. So I spent a couple of weeks with Gephardt, looked over the raw materials of his stump speech, including some trade statistics he'd gathered, and helped him frame the issue that would resonate most with voters: restoring competitiveness to American workers.

That day, when he got up to speak, he nailed the lines I'd given him, including a reference to the Hyundai, an import from Korea. And the room blew up. A huge ovation. Soon Shrum was in Iowa, making what became somewhat famously known as the "Hyundai" spot.[1] In it,

---

[1] The Hyundai spot became semi-famous among political consultants, in part because of the controversy over who came up with it first, Shrum or me. The truth is that it probably doesn't matter. I always thought that when our divergent energies clicked, Bob and I did our best work together.

Gephardt says he is "tired of hearing American workers blamed" for the trade imbalance with countries like Japan and Korea, and he shifts the blame from the workers to the administration's inability to stand up to Korea, where—because of taxes and tariffs—a $10,000 Chrysler car sold for $48,000.

With Gephardt as president, he said in the ad, the Koreans would "know that we'll still honor our treaties, because that's the kind of country we are. But they'd also be left asking themselves: How many Hyundais are they going to be able to sell in America for $48,000 apiece?"

After the ad began running, Gephardt finally caught with voters and he began to rise to the top of the polls in Iowa where he would eventually win.

This was what I did best, fine-tuning the candidate's strengths (and just as often, the opponent's weaknesses) into symbolic, blunt, and concrete messages that stuck with people.

And yet, from the beginning, there was something missing in this kind of politics for me. Even at its best, making TV spots can feel ultimately soulless. I'd spent nine years as a grunt organizer, from the Iola Williams bootstrap campaign in 1975 to Walter Mondale's establishment machine in 1984 and while the stress and physical toll were high, it was the thing I loved, the thing I was best at. I had passion for these candidates, and passion for organizing support for them. The people out there weren't simply raw numbers, or test cases performing behind a two-way mirror in a focus group. They were real people, in their homes, and that's where we engaged them—by phone or by letter, or best of all, by ringing their doorbells, by coming into their living rooms. And we didn't just try to spark an effect in them—to scare them or tap into some bias. We engaged them in the process, asked them to walk a precinct or make phone calls or put a sign up in a window. We *knew them*—the Jimmy Hogans and Irv Gadients of the world.

I was doing something I believed in, alongside people I believed in, for candidates I believed in.

But at Doak & Shrum, in 1986 alone, I worked on spots for more than twenty different Democratic campaigns. And while I agreed with most of the candidates whose ads I made, I didn't know what made them tick. Hell, *I hadn't even met* half of them. This wasn't something I really believed in anymore—like Iola Williams deserving a spot on the city council; or that Ted Kennedy would make a great president. I was selling something. Who it was didn't really matter. I may as well have been making commercials for mouthwash.

A television ad reaches voters passively. You just sit there and the box tells you what to think, or what you want—unlike the Internet, where you open a search engine or eBay or Amazon.com and *tell the box* what you want. And then, after you order the book, or bid on the baseball cards, the box asks you what you thought of the book and if you were happy with the auction. This is bottom-up, interactive communication. Television has a top-down, one-to-many structure, and it works by making an impression so that the next time you're in the grocery store and you walk past the Listerine, an image flashes in your mind of two actors so taken with each other's minty breath that they begin making out. It's no different with political advertising on TV: You sit there and the ad washes over you. If it's done well, some images stick, possibly even some ideas—although in a 30-second spot, there is usually only time for one or two visceral reactions to stick. Children playing: good. Pollution: bad. Old people in church: good. Criminals on street corners: bad. This is one of the reasons political debate in our country has been "dumbed down"—not because media consultants don't want you to think—but because no serious discussion of issues can possibly occur in 30 seconds.

If you make these short, reductive, cynical spots for a living, you begin to see the audience not as people organized into communities, with jobs and families and concerns, but as a series of effects you're trying to create, numbers to be moved, prejudices to be mined.

Occasionally, as a political media consultant, you might get to see real, live human beings in a focus group, but they're like rats in a lab—simply used to find the best way to identify and manipulate the real targets of the spot.

This reliance on polling and focus groups is—bar none—the worst trend in politics. It substitutes a candidate's convictions with bland, market-tested, centrist bullshit. If I could rub a lamp and get one wish to change politics, it wouldn't be to eliminate negative campaigning. I'd get rid of polling. I have nothing against pollsters. (What's the saying? Some of my best friends are pollsters.) But I have come to believe that polling takes much of the courage out of politics.

Faced with this kind of decaying politics, I'd love to say that I rebelled right away, that I woke up one morning and walked off into the sunset. Instead what I did during the late 1980s was become an expert at it.

VOTE FOR THE DEAD GUY

Negative campaigning has always existed. Always. That's the first thing you have to realize. In fact, today's most vicious attack ads are pats on the hand compared to the mud our forefathers slung.

In 1800, John Adams called his rival Thomas Jefferson a pagan, an atheist, and a traitor. His campaign said that if Jefferson were elected, "murder, robbery, rape, adultery, and incest will be openly taught and practiced." When that lovely bouquet didn't work (You have to feel for the political consultant assigned to go negative on Jefferson, to make *Soft on Crime* or *Has No Vision* stick to the man who wrote the frickin' Declaration of Independence), they went even lower and said the same dastardly thing that Republicans say about John Kerry today.

That he was too French.

Another popular negative tack at the time was to put some riders on horseback and send them around to villages spreading the rumor that your opponent (Jefferson) had died. (Today, we would test this in front of a focus group first—*Do you think you would be* more *or* less *likely to vote for a candidate who was dead?*) This left early campaigners with the task of sending out their own guys on horseback with the very modern problem of trying to disprove a negative: that the candidate was, in fact, alive.

No, the problem isn't that politicians in the twentieth century suddenly begun attacking their opponents. I would even argue that a candidate has a responsibility to the American people to tell us why the other guy *shouldn't* be president. That's not a problem.

The problem, again, is television.

Just to be clear, I'm not talking about the device itself, which is great. I have one. I watch Perry Mason reruns on it late at night. And *Law & Order*. And *The Sopranos*. And baseball. The problem is *the way it fails to engage people.*

In the last half of the twentieth century, television staged a hostile takeover of American culture, in just twenty years going from reflecting American life, to altering American life, to dictating nearly every aspect of American life: the products we buy, the clothes we wear, the things we fear.

Politics didn't escape the onslaught of TV. In some ways, it was one of TV's first conquests.

In 1948, Harry Truman's presidential campaign consisted of 31,000 different campaign stops, most of which involved him waving from the back of a train car. Let's say every time that train stopped, an average of six hundred people gathered to see him lean over the bunting and wave (it's an arbitrary number; some crowds were undoubtedly bigger, some smaller, and many people probably saw him more than once). But at six hundred people per stop, he might've reached nineteen million Americans.

Just four years later, in 1952, Dwight Eisenhower reached those same nineteen million Americans by simply driving to New York and standing in front of a TV camera for a few minutes.[2] Just in this one example, you see the roots of the insidious and destructive effect of TV on politics. Those nineteen million Americans were active and social in seeing Truman, going down to the train station, standing with other people, talking with them. The people who saw Eisenhower four years later did nothing but turn on the TV.

In 1948, when Truman was running for president, only 1 percent of Americans had a television set in their house. By 1956, the year I was born, when Ike was running for reelection, that number was 75 percent. By the time I began filming ads for candidates in 1985, the average American had *two* televisions in the house. In 1956, the average American watched about four hours of television. By 2000, that number was above seven hours a day.

As I've said, TV is a passive, top-down medium. Sitting around watching television inspires nothing but more sitting around and watching more television. As Robert Putnam writes in *Bowling Alone,* "television watching and especially dependence upon television for entertainment are closely correlated with civic disengagement." People who call TV their "primary form of entertainment" are far less likely to go to church, to write letters, to join clubs and organizations, or to attend public meetings. One estimate is that every hour of television watching translates to a 10 percent drop in civic involvement.[3]

---

[2] Museum of Radio and Television.

[3] Robert D. Putnam, *Bowling Alone: The Collapse and Revival of American Community* (New York: Simon & Schuster, 2000), Putnam's analysis of a Roper Social and Political Trends survey.

The modern political dependence on TV is usually traced to that moment in 1960 when Richard Nixon and John F. Kennedy debated on television, and Nixon's afternoon shadow and grim appearance made him seem untrustworthy while Kennedy seemed youthful and energetic. (In the long run, of course, Nixon *was* untrustworthy.)

I think the Eureka moment actually came eight years earlier, when Nixon was in danger of being dropped as Eisenhower's running mate over allegations that he'd been caught with his hand in the cookie jar—taking money and gifts from special interests in exchange for influence. (Look how far we've come; under George W. Bush, this is the *definition* of government.) In 1952, Nixon responded with the sticky-sweet and self-pitying "Checkers speech," in which he deflected the criticism of the money he'd taken by defiantly focusing on a single gift from a Texas supporter:

> It was a little cocker spaniel dog in a crate that he'd sent all the way from Texas. Black-and-white spotted. And our little girl—Tricia, the six-year-old—named it Checkers. And you know, the kids, like all kids, love the dog and I just want to say this right now, that regardless of what they say about it, we're gonna keep it.

Just like that, bullshit had its own medium. And far worse was to come.

Now, fifty years of this stuff has so fractured and dulled our senses—even as we have more and more competing channels and more ways to skip around when we're bored or when we know we're being lied to—that each year it takes more to catch the scattered attention of viewers.

This is a problem for politics, just like it's a problem for advertisers of every manner of product. Each year, TV advertising becomes more dominant, more ubiquitous, and more expensive, at the same time it becomes *less effective*.

This is why, in politics, we've become caught in a self-defeating cycle.

First, you have to understand: negative ads work. Perhaps it's human nature, but if you believe that Thomas Jefferson is a traitor, or that Michael Dukakis is a wimp, or that George W. Bush is a blithering idiot, you are less likely to vote for him.

The problem is that as TV inexorably decreased the attention span of Americans, political consultants realized that they had no choice but to go with the one most effective ad, the one that "sticks." This created the downward cycle in our political process because negative ads cause people to react viscerally, not just to the person they're aimed at, but to politicians as a whole.

This is why broadcast politics has failed us so miserably. Not because it doesn't work, but because the most effective ads are the one that make the community a worse place to live. Think about it: *the best ads—deceitful, negative, and manipulative—are the ones that fail the country the most.*

And so, as people become more cynical about politics because of this manipulation and negativity, they become naturally less engaged and inquisitive. For the political operatives, the only way to reach this disengaged electorate is to sharpen the ads to an even finer point, to go even *more negative.* It's like shining a bright light in someone's eye—the pupil closes the brighter the light gets, letting in less light, so you make it brighter to get more light in and the pupil shuts out more light, until pretty soon, the person just has to look away.

When the topic was politics in the late twentieth century, Americans increasingly looked away. They had better things to do with their lives.

If TV advertising worked in any kind of positive way, then it would follow that more people would engage in the process as TV political advertising increased. But the last real increase in voter participation was in the 1950s and early 1960s, when Jim Crow laws were being overturned and African Americans finally were getting the opportunity to go to the polls in the South. Yet between 1960 and now, the percentage of voting age adults who cast their ballots for president has fallen by about 15 percent while the amount of money spent on campaign advertising increased exponentially from a few million to the *$1.6 billion* that's expected to be thrown at the 2004 election. The percentage of people who voted in congressional races fell even further, from about 30 percent in 1970 to about 15 percent now.[4] That is not a very good return on investment. At the same time, news programs were devoting less and less time to politics. A 2000 Harvard Kennedy School of Government study found that news "without a public policy component" rose from about 30 percent of the newscast in 1980 to almost 50 percent in the year 2000. In just one eight-year news

---

[4] Thomas E. Patterson, *The Vanishing Voter* (New York: Alfred A. Knopf, 2002).

period, 1994 to 2002, the amount of comparable coverage of political campaigns on the networks fell by 72 percent.[5] People were four times as likely to get their information about a candidate from campaign ads than from newscasts.[6]

The rock-bottom low point in broadcast politics came in June of 1988, when Michael Dukakis clinched the Democratic nomination and looked like he had a good shot to beat then Vice President George H. W. Bush. With his primary bump, Dukakis led in the polls 52 to 38. But any pollster worth his focus group will tell you that a more telling number was the candidates' Favorable/Unfavorable rating.

It's a problem for incumbents everywhere when they've been in office long enough for people to genuinely hate them. So in June, while 40 percent of Americans had an "unfavorable" reaction to George Bush, only 20 percent had the same gut feeling about Duke. Bush's "favorable" rating, meanwhile was 53 percent, Dukakis's a whopping 70.

Obviously, Bush's guys could have tried to improve their candidate's favorable rating, but it was far easier in a thirty-second TV spot to lower Dukakis's. Each campaign does "opposition research" against its opponents and when the time is right, dumps the "oppo file" on reporters and in attack ads, hoping to define the other guy out of the race (this is the current Bush's strategy against John Kerry). In 1988, Bush's key strategist Lee Atwater—and a supporting army of conservative political action groups—launched a hailstorm of negative TV ads focused on two visceral images: a picture of the former Massachusetts governor sitting in a tank, looking goofy (Dufus in tank=Soft on military) and even more cynical and more effective: the image of Dukakis letting murderers go home for the weekend as part of an experimental prison furlough plan (Dukakis lets Black men leave jail early = Fear for your lives).

These ads were deadly effective. The most famous, of course, was the Willie Horton spot, a minimalist attack masterpiece, paid for by the National Security PAC. It reported that as governor, Dukakis allowed "murderers to have weekend passes" and then flashed the mugshot of a black man named Willie Horton, who had committed heinous crimes

---

[5] University of Southern California's Annenberg School and the University of Wisconsin Advertising Project.
[6] The Center for Media and Public Affairs.

while on furlough. From the picture of Horton, the words: "Kidnapping, Stabbing, Raping" bled into another picture of Dukakis.

Another ad, this one produced by the Bush campaign, reported that Dukakis gave furloughs to "first-degree murderers not eligible for parole," that "268 escaped" and "many are still at large." In fact, those escapes were over a ten-year period and half of them hadn't escaped but simply returned late. Only three were actually at large. And only four of the 268 "escapees" were murderers not eligible for parole.

The media reported on both the racist tone and the factual errors in the ads, but the reporting only increased their penetration and in the end, these visceral missiles hit their target. By July 8, 1988, just a month after scoring a 70 favorable rating, and *two weeks* into the Bush campaign attack on the furlough program, the Massachusetts governor's favorable rating had fallen 13 points to 57, and his lead in the polls had narrowed to 6 points. Later, the Bush team added a spot that showed homeless people dressed up as prisoners leaving prison through a revolving door. Bush strategist Lee Atwater was making good on his promise to "strip the bark off the little bastard" Dukakis and "make Willie Horton his running mate."[7]

By election day, Dukakis's favorable and unfavorable numbers were exactly equal: 45 percent liked him and 45 percent didn't. Bush's stayed pretty much the same, and he cruised to victory not because he convinced more people that he should be president, but because he scared more people into believing that his opponent might release criminals into the streets—something that any president would be hard pressed to do even if he was so inclined. By the time the election was over, *70 percent of Americans* wished they had another person to vote for. This is what the system brings us: negative ads so effective that when the campaign was over, Americans hated *both candidates.*

The 1988 presidential election was a sea change. If you look at Ronald Reagan's slick spots four years earlier, you don't see anything that approaches the vitriol and ruthlessness of those 1988 ads. What followed in each election since then—presidential and nonpresidential—was an all-out war, a bludgeoning mentality, the use of a blunt force instrument to take out the opponent.

---

[7] Charles Lewis and the Center for Public Integrity, *The Buying of the President 2004* (New York: HarperCollins, 2004).

We are left with a system of mutually assured destruction, both parties constantly lobbing nuclear missiles at one another—scorched earth. And what gets destroyed is our democracy. What is defeated is the people's desire to get involved.

DISILLUSIONMENT AND DEATH

Gary Hart seemed like the real thing, although in a totally different way than Walter Mondale. Brilliant and handsome, Hart had thought more about terrorism in 1987 than most candidates have today. He absolutely recognized the problems that America was going to face and was calling for a smaller, more agile military to deal with the realities of a world where the real threat was about to come not from the old Soviet Union but from the instability of terrorist groups and the third world.

As a candidate, he could be aloof and arrogant, but he also had the goods—that rare Clintonian combination of substance and appeal—an unrivaled knowledge of world affairs to go with great hair.

Of course, I had vowed to never, ever work day-to-day on another presidential campaign, but I also had to get away from the deadening, assembly line work of churning out political spots for candidates I didn't know. I wanted to return to the kind of politics—and the kind of candidate—that I had once been passionate about. I was also intrigued by Hart's concentric circle campaign of 1984, and since Tully had gone to work for Hart, I ignored my own good sense and, in May 1987, followed him into battle, signing on as the campaign's deputy political director.

But just a few weeks after I'd started, Hart, in the words of one observer, "shot himself in the testicles." Asked about rumors of his infidelity he told a reporter from the *New York Times* magazine that the media should "follow me around. I don't care . . . if anybody wants to put a tail on me, go ahead. They'd be very bored."[8]

The very weekend that story was scheduled to run, a *Miami Herald* reporter was indeed staking him out and spotted Hart with a twenty-nine-year-old actress named Donna Rice. Soon a photo of the two of them taken a month earlier was circulating: the blond on his lap in front

---

[8] Peter Goldman, Thomas M. DeFrank, Mark Miller, Andrew Murr, and Tom Mathews, *Quest for the Presidency, 1992* (College Station, TX: Texas A&M Press, 1993).

of the aptly-named yacht, *Monkey Business*. Within days, he'd dropped out of the race.

I believed then, like now, that a candidate's sex life has absolutely nothing to do with his ability to be president. Obviously, you can question the man's judgment—knowing that his private life is under intense scrutiny and daring reporters to follow him. But I also wonder about *our* judgment, the judgment of those in the sensationalist media and all of us who eagerly consume it. Think of what it cost us, to lose a man who might have been a great president, a man who knew more twenty years ago about the world we would face than many of the candidates running now. And for what: a tittering, meaningless story about what happened between two adults. A transgression that Hart's own wife had forgiven him for. Broadcast politics was more interested in tailing Gary Hart than in reporting the farsighted positions he staked out nearly twenty years ahead of his time. In the wake of September 11, 2001, which mistake was bigger? His? Or ours?

I've worked for seven presidential candidates—they all have flaws—and broadcast politics seems to exist for the sole purpose of finding them. So we throw out one of the leading thinkers in our nation on terrorism because he's not perfect? Where does this end, when you throw out candidates because they're *too human,* because they have trouble in their marriages, because they yell in their concession speeches?

Unfortunately, Gary Hart's trouble was just beginning the country's ridiculous obsession with the private lives of our politicians.

Suddenly without a candidate and without a job, I landed as deputy campaign manager in charge of message and media for the Gephardt campaign, which was being run by my old pal from the Kennedy days, Bill Carrick. Steve Murphy and Mike Ford were also working on it, so it was natural to approach the 1988 campaign with the same gonzo fervor I'd brought to the 1980 and 1984 races. It's the only way I know to campaign. In Iowa, I even tried to get Gephardt to do a campaign stop at Irv Gadient's barn, but he had died since the last caucuses. National politics was dying inside me, too. After the past few years of making empty, cynical TV ads, after message-testing, flip-flopping candidates, after Gary Hart's implosion, the whole system seemed terminally ill to me. I genuinely liked and respected Dick Gephardt, but I didn't know if I could stomach the system anymore.

We built enough momentum after the Hyundai ad to win Iowa, but the campaign shot its wad there and in New Hampshire, and we got almost no boost from it. Out of money, Gephardt began flaming out when the Dukakis campaign, flush with cash, ran ad after ad of a Dick Gephardt look-a-like complete in a gymnast's outfit doing massive summersaults and triple flips on a trampoline to an announcer's dramatic recital of Gephardt's record as a flip-flopper, abuser of gravity, and the laws of physics and everything known to be good in America.

When the race was over, I swore on my life I'd never to do another presidential. I was beaten, bitter, and broke (I still have never been reimbursed for $89,000 in campaign expenses). I had gotten married in 1984 and my wife Katie and I had begun having children—Christine was two in 1988; Jim and Ted would come along later—and the eighteen-hour days and seven-day weeks required to lead a campaign life simply didn't fit with an adult trying to raise a family.

Of course, I had sworn off presidential politics before. And as disillusioned as I was in 1988, I suppose that I might have been lulled back by the right candidate. But a series of events convinced me that this time, I really was done—at least with day-to-day campaigning.

A few years earlier, when I was applying for health insurance as part of my employment with Doak and Shrum I got one of those rejection notices saying that something was wrong with my health. It turned out my blood sugar was dangerously high. I had adult onset diabetes. This is common among pols, caused by stress, poor diet, and lack of rest (pretty much the job description for a campaign staffer). I could manage it, the doctors said, by changing my diet, getting more sleep, and—most of all—staying away from campaigns.

It hadn't been enough to keep me out of the 1988 scrum. But there was more reality coming.

Over the years, my father and I had begun a slow reconciliation. On the road in 1984, I'd shared with Mondale how much my dad hated that I had gone off to be a "political hack." The night of the Pennsylvania primary—after we'd gone from 14 points down to a 14-point win—I went to get Mondale for the victory celebration and he was standing in the hotel room next to a scrawny old Italian guy. My dad.

Fritz had flown him to Philadelphia.

Mondale's comeback had started with a pair of symbolic boxing gloves and, in front of my dad, on the night of my biggest win of that campaign, he signed them—"To Rocky Trippi"—and handed them to me. And then he told my dad, "I wouldn't be here if it weren't for your son."

By 1988, Dad and I were even talking about taking a trip together to Sicily, to see the old country. After Gephardt was knocked out, I took a Senate campaign in Wyoming and I told my father we'd leave for Italy the day after the election.

We had our tickets and were all set to go, but the Wyoming race turned out to be too close to call and I was sent to Casper to oversee the recount for the Democrats. I told my dad it would only be a short delay, maybe a few days, and then we'd take our trip.

When I landed, it seemed like everyone was looking for me. They led me to a wall where someone was standing, holding a pay phone. The voice on the other end had a Kentucky lilt: "Joe, I don't know how to tell you this. But your daddy passed away."

He'd died of a heart attack. I looked around the airport. I had given up a trip with my father to his homeland for this: to recount votes for some fucking election in Wyoming?

At his funeral, I laid one of Mondale's red boxing gloves in the casket with him.

These were hard years for the tribe of campaigners I'd worked with, too.

In 1988, Tony Pappas, whom I'd met my first day in Iowa working for Ted Kennedy, and who had been working as a congressional staffer, committed suicide by jumping out the window of an eighteen-story building the day a *Washington Post* story repeated rumors that he was gay.

Then, in September 1992, the indestructible Paul Tully died suddenly of a heart attack while working in Little Rock, Arkansas, on Bill Clinton's campaign. We had all worried about him—his size, his health, his drive, and his diet—but you never think it's going to happen. It was impossible to imagine a world without Tully.

My own life had taken a beating in those years, too. I was working hard to adjust my lifestyle, trying to keep my diabetes in check. My marriage had dissolved. And then, in 1994, I was shooting an ad for a gubernatorial race when I got a telephone call saying that my five-year-old son,

Teddy, had been hit by a car and had severe leg and head injuries and would probably not live.

I flew home and Ted's mother, Katie, and I took turns keeping vigil at Ted's bedside. He was in a full body cast for months and, because pain relievers could potentially induce a coma, was given psychotropic drugs. It was torture on both Katie and me, watching a five-year-old struggle with the pain and disorientation caused by the medicines they gave him. Then, one day, we noticed that something about him wasn't right; he seemed off. The doctors took him in for an MRI scan and, to this day, I can still see the looks of shock on their faces. They turned off the machine and rushed him into surgery. His brain was hemorrhaging right at that moment. They literally watched it erupt under the MRI scan. He was in surgery for thirteen hours. But somehow they managed to save his life.

Ted's accident was the low point of those tough years, and I think it helped put politics in perspective for me. Here I was, divorced, my father was gone, dealing with diabetes and my son's long, slow recovery. I couldn't imagine that I would ever invest as much of myself in a campaign as I had in the early 1980s.

In 1988, I started my own media company, Trippi and Associates, in part so that I could be more selective about the races that I handled. But I was late getting into the 1989 races and all my old spots were the property of Doak and Shrum. So I had no reel—no sample of my work to show potential clients. I found myself cold-calling candidates, turning a light on a wall and doing little ads with shadow puppets.

Needless to say, some people didn't get it—and who could blame them?

But when I did my little shadow dance for the mayor of Alexandria, Virginia, Jim Moran, who was making a run for Congress, he hired me on the spot. His formidable campaign manager, Mame Reiley, took a little more convincing that the shadow puppet guy was a good fit, but I quickly won her over, along with her assistant, a bright beautiful blond livewire named Kathy Lash.

I became great friends with the eternally sunny Kathy and with Mame, who is nothing less than a force of nature herself. We were the Three Musketeers, rolling around town, meeting every Tuesday for a drink to talk politics. After we'd been friends for years, one day in the mid-1990s Mame suggested that I ask Kathy out. I thought she had finally lost it. There was no way this amazing woman would date me. But she did, and in 1997, we were married.

In the 1990s, for the first time, my life felt balanced. I was spending more time working on technology issues, picking and choosing the races I worked on, only accepting candidates I believed in. I kept my vow to stay away from day-to-day presidential campaigns, but in 1992, got an intriguing call from Virginia Governor Doug Wilder.

An up-by-his-bootstraps, Korean War hero who had erased a $2 billion budget deficit in Virginia without raising taxes, Wilder was a natural. He was also African American.

But he didn't want to run his campaign as a reflection of the struggle for civil rights in America, the way Jesse Jackson had so effectively campaigned. Not that he would ignore his ethnicity, but it would simply be a part of the man—not the sum. This was the kind of campaign that I believed in passionately, and I agreed to do some limited consulting for him.

Unfortunately, the focus groups we gathered wouldn't let Doug Wilder's race be simply a part of the man. When we laid out his story next to the other candidates—without their pictures or names, just their general biographies—Wilder shot to the top. But when we showed these people a television ad that began with Wilder's picture, almost all of them withdrew their support. Some people reacted out of pure bigotry, others a frustrating kind of pragmatism: *I just don't think he could ever get elected.*

It was disappointing that in 1991 this was where we stood as a country, that in many ways it hadn't progressed any further than the country Bobby Kennedy had tried to heal back in 1968.

A year earlier, I had begun to wonder if it wasn't possible to use television a different way. Couldn't you make it interactive, use it to inspire people to *do something,* instead of just bombarding them with American flags or mug shots of criminals. If this thing was powerful enough to nearly destroy participatory democracy, it ought to be powerful enough to rebuild it.

Just before I signed on with Wilder, I had gone to a casual dinner at a Mexican restaurant in Washington, D.C., with Governor Jerry Brown, Mike Ford, and some other people. While the political establishment viewed Brown as the eccentric Governor Moonbeam, Mike and I had always had a great deal of respect for him. We knew that Brown was considering a run for the White House and we were encouraging him to do it. One of the things we talked about that night was the possibility of using a toll-free 800 number to attract voters.

Soon after that, I signed on to do some limited consulting for Wilder and Ford and ended up serving the same emeritus role for Brown.

So you can imagine my chagrin at the first debate in December, as Wilder more than held his own, but was completely overshadowed by Jerry Brown, who answered every question by holding up a sign with his 800 number on it. Brokaw asked Brown if he could please refrain from holding up the number when he answered questions, and Brown answered that question by holding up the sign. "No, Tom. I can't . . ."[9]

At Brown campaign headquarters, supporters and money were rolling in, even though Brown had announced that he wouldn't take any donation larger than $100, one of those counter-intuitive ideas that I filed away and would come back to years later. At the commercial break, a frustrated Brokaw reached for the gas can to put out the 800-number fire, calling Mike Ford and asking him to talk some sense into his candidate. Sure, Mike said, and then he sat back and watched as that 800 number became the key to Brown's surprising run—his message, his organization, his fund-raising all rolled into one media-savvy, interactive bundle.

Later, I found out that at about the same time I was tossing around the idea of an 800 number for a politician, Brown's brilliant, young campaign strategist Joe Costello was having the same epiphany, watching televange-lists and those late-night infomercials in which they flashed a toll-free 800 number and bugged people to call in.

In the end, when all the other Democrats in a crowded field had dropped out, only Brown would still be standing, nudging the party to the left against the best campaigner of our generation, Bill Clinton, and with $100 donations on an 800 number, reminding people that this was *their* system, not the Democrats' or the Republicans' or the special interests'. It was theirs.

Joe Costello kept coming up with ingenious new ways to plug the number and when Wilder dropped out of the race, my partner Mark Squier and I stepped in to make Brown's "We the People" TV spots, which pushed for the idea of empowering people to take action for change, and which

[9] Peter Goldman, Thomas M. DeFrank, Mark Miller, Andrew Murr, and Tom Mathews, *Quest for the Presidency, 1992* (College Station, TX: Texas A&M Press, 1993).

were built around the 800 number and the idea of getting Americans involved again. The Brown 800 number raised about $5 million in small donations. I never forgot the lessons of that campaign.

Looking back, it's easy to see it as the first step in many that would lead to the Dean campaign, but it's amazing how primitive it was—like one of those old washing-machine sized computers, big, bulky, and not quite there yet . . . but in hindsight, the big, messy beginnings of a revolution.

4

THE MOMENT

Killer Apps, Open Source, and The Guy

YOU JUST DON'T get it.

If you polled the staffers of Dean for America and asked them what is the first thing that pops in your head when you hear the name Joe Trippi, an overwhelming majority would start with that phrase. It was the chorus of the song I went around belting at the top of my lungs for a year—that the leaders of American politics, media, entertainment, and business were in the dark about what we were doing in the Dean campaign, that they were clinging to old methods and ideas that were about to become archaeological relics right before their fossilized, corporate, country-club, never-gonna-get-it nearsighted eyes.

The media? The party leaders? The other campaigns? *Don't frickin' get it.*

And if I said this to your face . . . I could call you a son-of-a-bitch ass-hole moron and mean it endearingly, but when I looked you in the eye and said, "You just don't get it . . ." watch out.

It's always been one of the hardest things for me, and one of my chief frustrations with myself—this difficulty I sometimes have explaining things that seem obvious to me. With politics, for instance, I can look at a candidate, his message, the field, and the issues and see in an instant where the potential holes will be in the campaign. Speeches, slogans, whole TV spots—pictures and words—come screaming into my head. I have absolutely no idea where they come from and when I sit down and transcribe them onto paper and put a stop watch on them, they are exactly the length of a TV spot: 30 seconds. I'm the first to admit, it's a little bit weird. Even scary. My working style—late-night pacing distracted manic hand-waving—can be jarring for someone who is more used to . . . say . . . desks. Or sleep. So I understand people who describe me as a mad scientist. Sometimes when I'm explaining how to patch the holes, the person I'm talking to hasn't seen the holes yet and can't see the pictures and words in my head—*hell, I didn't see it myself until a few seconds earlier*—and so my patches seem arbitrary, as if I'm frantically trying to fix a hull that isn't leaking yet. When this happens, I can come across as short with people—scratch that: I am short with people—dismissive, or just plain nuts.

My reaction is sometimes the same to people who don't see the impacts that the Internet and other technology are already having on our lives. There has always been a disconnect in America between those people who look at a computer screen and see what is, and those people who look at the same screen and see *what's going to be,* between those people who know the world is changing profoundly before our eyes and those people who—for one reason or another—just don't get it.

I still remember the precise moment *I* got it.

Throughout the 1990s, I had drifted slowly away from politics and into the other world in which I moved comfortably—technology. It wasn't that my passion for transforming the world had dried up. If anything, it was stronger than ever. But the change in politics worked at a snail's pace compared to the world of computers and telecommunications, which were delivering *now* on the promise of giving people more knowledge, and therefore more power over their own lives.

Whatever talent I had in politics—mostly in helping candidates communicate their views and ideas about the specialized world of government—was also in demand in the "new economy," this river of technology and communications companies that had spilled over its banks in Seattle and the Silicon Valley and was flooding the entire country with innovation.

And money.

In the late 1990s, this was the thing that impressed most people about businesses like Amazon.com, Intel, and Lucent Technologies—their price per share.

Not that I didn't care about the stock market, but when I began consulting with high-tech startups, I found myself choosing clients the same way I chose political races, a calculation that Paul Tully used to describe as "finding the most progressive candidate who actually has a prayer of winning."

I would become fascinated by the newest, most speculative, and futuristic high-tech concept, seek out the business that was doing the best job working on problems associated with that idea, and offer my consulting services to help translate and refine the "message" of their technology, on the off chance we could get it included in the newest wave of personal computers (this is about as easy as getting someone elected president). Even in political years, when I was immersed in campaigns and had to keep one eye on *Newsweek,* the other eye was always on *Wired.*

Pols tend to be a single-minded group during election season. Open a campaign guy's briefcase and you're likely to find polling numbers, message memos, and news magazines. So I was easy to spot, the one in the airport with my nose in a textbook about multiwavelength optical networks or flipping through an article on the ramifications of infinite bandwidth. I still threw everything I had into the campaigns that I worked on, but it wasn't the only thing I did anymore. I'd finish a meeting with a U.S. Senator about television spots for his upcoming race, walk five blocks to a coffee shop and sit down with a couple of twenty-six-year-olds to talk about why a chip on the southbridge of the motherboard with triple-DES encryption was the best personal computing security solution. More than ever, I had a foot in these two worlds, with this sneaking suspicion that the gap between them was narrowing every day.

The late 1990s were thrilling. The rate of technological change in America was increasing exponentially. Fueled by happy investors, measures like processing speed and chip memory doubled themselves in increasingly shorter periods of time, allowing for even faster developments elsewhere. New innovations shot up everywhere and everyone was looking for that killer application, the gap constantly narrowing between science fiction and reality, between imagination and innovation.

For that brief, wonderful period, daydreams became a viable commodity. One of my two lifelong visions was coming true: we were living in a geek-tech wonderland. The only thing that would have made me happier was—in the midst of this technological renaissance—to have a Bobby Kennedy-style candidate emerge, reenergize people again, and begin re-making the country.

But you can't have everything. In spite of Bill Clinton's popularity, the Republicans had a chokehold on both houses of Congress. So, during election years, I continued to work half-time trying to get Democrats elected, but in between campaigns, the vast majority of my time was spent working for a handful of companies, mostly startups, in whom I saw the same insurgent, progressive, bottom-up, long-shot, democratic philosophy that I looked for in candidates.

This led me to do consulting work for a few brash young companies:

- Wave Systems, which was developing technology to create trust at the edge of the Internet, helping people better control the security and dissemination of video and other content on their own computers (one upshot being that you could post your own music videos, or your secret recipe for chili online and charge a nickel every time someone downloaded it).
- Progeny Linux Systems, which was writing software for a version of Linux, the increasingly popular, stable, bottom-up, and most important—free—operating system (think of it as the anti-Windows) that had taken the original intent of the Internet and run with it, creating an open-source computer code that invited innovation instead of choking it.
- Smart Paper Networks, which was exactly what it sounded like—an interactive kind of paper that reacted when you touched it by linking up the reader with web sites related to the thing you'd touched (so that, as you read a magazine article that mentioned the music of Bob Marley, your computer screen could jump to a video of Marley performing, a catalogue of his songs or a store where you could order his music. I was amazed by this then, and I still am).

For me, the most promising thing about the Internet in those days was the way it transformed communication, the way it actually reversed some of the more insidious aspects of television. It was making people *talk to each other* again.

I certainly did my share. I visited chat rooms and posted messages all over the web on electronic bulletin boards about investing, politics, technology, sports. One of the most amazing online communities formed around Wave Systems, in a chat room for investors, people called *Wavoids* (my on-screen name was "random1"). The amazing thing was that the company's leaders actually read what their investors had to say. They actually listened. But they didn't communicate as well as they could have with this community of investors—at least not in English. So I wrote the CEO and soon I was the one translating the highly technical world of Wave for lay people. At its peak, there were more than 1,000 posts a day on the Wave bulletin board. I also contributed to the raucous debates on Raging Bull and Motley Fool, and I can proudly say that I was once poster of the week on a board devoted to Baltimore Orioles fans.

In the late 1990s, I spent a lot of time on web sites reading about companies I'd invested in, like one company called THQ, which made games for PlayStation and GameBoy and other video game consoles. It was a small tech company that had grown successful and there was a happy club of investors, game players, and the curious who gathered on the web to read about what was happening and comment about the company's new games, its developing technology and, generally, what it was up to.

This young day trader named David Haines (he posted as HAINESDA) was the unofficial star of the THQ bulletin board. Everyone came to see what David would write. He was twenty-nine or thirty, funny, articulate, informative, and if you followed David, you got a real report on the company—which games rocked, which killer apps lay just around the corner, what the company was up to—as opposed to the rows of cold numbers and corporate jargon in the quarterly reports that traditional companies send out to their shareholders. And unlike a corporate report, you could actually talk back to David and he'd answer your questions, comment on your idea, laugh at your jokes. At one point, a group of online investors even chipped in four hundred bucks to send David to a gaming convention because they couldn't imagine not having David's take on it. At the convention, the THQ engineers took David on a guided tour, treating him not like a small investor, but like a partner in their innovation. This was a whole new business model, the company sharing information with a key person, who shared it with the people who use it and invest in it, unlike the old businesses, which viewed its customers the way the military viewed targets, making secret marketing plans and then "unveiling" them on the public with a manipulative advertising blitz.

In the truest sense of bottom-up, companies like THQ responded to input from the people who actually *used* their stuff and invested in their company, on the off chance that the 200 brightest people in the world might not all work at their company, and that the people who played their games and bought their stock might not be total morons. Because of this, THQ didn't just have people who used the products—like the people who buy Bounty paper towels or who eat gorditas at Taco Bell—they had a *community,* thousands of people who cared enough to get involved with THQ, to engage in dialogues over the Internet, like people at an Iowa caucus whose neighborhood was not a row of farmhouses, but a row of web sites.

But more than just getting to know the company, readers of David's posts on the bulletin board *got to know David.* It turned out he was a tall, wiry triathlete with a cat named Sierra and a dog named Bruin. Once, when he missed a couple of posts, he got on to apologize, to say that his wife had just given birth to a baby boy, Christian. I never met David Haines and yet I *knew him.* For two years, I checked in with him once a week, sometimes more. I was there when his kids were born, when they walked for the first time.

And I was there the day someone else got on the board to report that David had died the day before of a heart attack. For days, the whole board—built around a gaming technology company of all things—went into a deep period of mourning. I sat at my PC, crying and watching as people eulogized David and mourned him the way you'd mourn a good friend.

And that is the precise moment that I *got it.*

I was attending a funeral on the Internet.

This was not a bunch of individual people sitting in front of a television alone, watching a sad program, reaching on cue for the Kleenex brand tissue. This was a rich, fully realized community, a world of real people interacting with each other, sharing their kids' first steps and crying on each other's shoulders when they lost someone they cared about, someone most of us had never met.

You get used to things moving in a certain direction. All the inertia carrying us downward, toward the place that society has fallen: a dark, cynical, self-defeating cave. And then you have this moment of hope. This moment in which you realize that you might be standing at the dawn of something profound, something that can reverse this descent into a depersonalized and disengaged mass. At keyboards all over the world, there could be a million, two million, *ten million* David Haineses, smart, funny, good people who want to change the world and have just been waiting for that moment when someone figures out how to get them all together.

AT THE INTERSECTION OF POLITICS AND TECHNOLOGY

So it was just plain dumb luck. Or maybe it was *crazy* plain dumb luck. Maybe the same congenital flaw that made me daft enough to work on six presidential campaigns was why I was constantly babbling in front of a computer screen, surfing Internet chat rooms, company bulletin boards, commenting on out-of-the-way blogs. Whatever the reason, I unwittingly found myself in the right spot at the beginning of the twenty-first century, maybe the one guy who might have a chance at pulling together this grassroots, nuts-and-bolts, knock-on-doors, send-e-mails, use-the-Internet-bulletin board-web site political movement *thing,*

As the elections of 1998, 2000, and 2002 slime-trailed their way across American TV screens—the moneyed congressional incumbents winning 98 percent of the races against challengers[1]—this idea was germinating in my scattered head. I bored my share of people in both worlds—politics and technology—with my riff about how one day in the very near future we'd build huge, involved communities around political issues and candidates, how these people would be an army, ready to mobilize at the first sign that the government was doing that top-down, trust-us-we-know-what's-best-for-you crap that people were so sick of. Government, the entertainment industry, and corporate America better get ready. The American people are going to learn how to organize themselves and then watch out.

At some point, the listener's eyes would glaze over and that person would nod politely, or sometimes not so politely.

You just don't get it.

---

[1] Charles Lewis and the Center for Public Integrity, *The Buying of the President 2004* (New York: HarperCollins, 2004).

No, the person would say. I get it.

No. You don't. You just don't get it.

When the tech stock bubble burst in 2000, the old guard of business and culture in America thought it had gotten a reprieve, that all that futuristic, utopian, techno hoo-haw they'd been hearing and mostly ignoring had turned out to be crap.

All together: *You don't get it.*

It took the telephone sixty-seven years to go from 1 percent of households to 75 percent. Took the car fifty-two years. The radio fourteen. The fastest innovation to get into 75 percent of households was the television, which took seven years. And that's about how long it took 75 percent of Americans to get on the Internet. Seven years.[2]

But this number is a little bit deceiving. When the TV got into the house, all you had to do was turn the thing on, and bang, Dwight Eisenhower was standing in your living room, warning you about the Red Menace. Everyone knew what a TV did. But for the Internet to really catch on—to become this force that could change the world—a few more things had to happen. Some snow had to be plowed.

First, Amazon.com had to show you that you could buy something on the Internet, in this case a book, and the damn thing would actually show up. That your credit card number wouldn't immediately be e-mailed to every criminal in the country. Sites like eBay had to show that an Internet communities could be a civil, orderly, law-abiding, welcoming place. In 1997, eBay reported that only 27 of its 2 million online auctions involved possible fraud. That's .01 percent, meaning 99.99 percent of transactions were on the level.[3] Find me a real city with that kind of crime rate and I'll move tomorrow. Or online travel. Early on, people would find the flight they wanted on the Internet, but would then call the airline rather than take the chance. Sites like Expedia and Travelocity had to show people that it was faster, easier, and usually cheaper to go online and make your own travel arrangements, and that you didn't need paper to get on a plane.

What these sites and others were doing was plowing snow, clearing the roads for Internet users to feel safe and comfortable enough to spend money,

[2] Robert D. Putnam, *Bowling Alone: The Collapse and Revival of American Community* (New York: Simon & Schuster, 2000).

[3] Howard Rheingold, *Smart Mobs* (Cambridge, MA: Perseus Books, 2002).

make reservations, and interact on the web in a million ways. While corporate leaders and stock analysts continued to miss the point by focusing on the rise and fall of admittedly inflated stock prices, the truth is that not long after the bubble burst—a couple of months maybe—more people were doing more things on the Internet than ever. If you look at the penetration of the telephone and the radio into American homes, it's a steady uphill climb—except for a few years during the Great Depression—when the number of phones and radios took a small dip. If there was a similar downtick with the Internet and the bursting of the stock bubble, it was even more short-lived. Now imagine if you owned a radio station in the early days of radio, in those fourteen years before 75 percent of Americans had a radio in their houses. What if, after ten years of losing money, you sold that station?

So go ahead and pretend the Internet died with the stock market bubble. In the words of Clint Eastwood, *Do you feel lucky? Well, do ya?*

Of course, there's a fine but excruciating line between being the first and being the *first to succeed*. There can be a ten- or fifteen-year lag time between the visionary and the vision. So for the first fifteen years, the naysayers win, and just before the thing is about to become a reality, a whole bunch of people give up on it. It's like that brick-size Apple Newton that I had in 1993, a failure in the business world, but the precursor to all of today's PDAs. In 2000, as tech companies lost their pipeline to easy funding, a bunch of great ideas went under, devices that, in ten years, we won't know how we lived without, services that will make so much sense that we can't imagine they haven't always been around. Like buying airline tickets on the Internet. Start up an online travel agency ten years ago and you're a bankrupt idiot. Start up an online travel agency two years ago and you're a billionaire.

It's no different with politics. I had watched the "1-800" populist candidates, like Jerry Brown and Ross Perot, and I closely followed John McCain's insurgent Republican presidential bid in 2000, the first national campaign to attempt to make use of the Internet. I held my breath that year—excited that someone was trying it, but terrified that they'd pull it off before I got the chance. They didn't. McCain managed to pull a decent number of people, about 40,000, into his campaign via the Internet, but it was the Newton of online political campaigns. The technology simply wasn't quite mature enough yet; enough snow hadn't been plowed.

As I daydreamed about how this thing might work, I thought the bottom-up, interactive Internet campaign that I was visualizing might still be eight or ten years away from working in politics. And it wasn't just a matter of plowing snow, or of getting all the technology in order, the way you would if you were developing a new chip or a lighter laptop.

No, to build *this startup,* first you needed to go out and find the guy.

A SURE LOSER

I was dead. Midway through the 2002 election cycle, I was as tired as I'd ever been in my life. I was sick. I was intending to sleep through 2003, sleep through the 2004 presidential, maybe wake up sometime in the fall of 2005, just in time to see the Orioles win the World Series.

Instead, I made the mistake of taking a campaign. And not just any campaign, but a knock-down, drag-out donnybrook, one of the hardest, most negative, most hellacious campaigns of my life.

It was the 2002 mid-term congressional elections, and Karl Rove and the Republicans had decided to cement their hold on Congress by taking advantage of the once-a-decade reapportionment of congressional districts. What they did was redraw the boundaries in five districts, so that five Democratic members of the House of Representatives woke up one morning to find themselves living in a mostly Republican district, the boundaries of which now extended like comic-book thought bubbles around their houses.

In these five Republican districts, there were now two incumbents: one displaced Democrat and one heavily favored Republican. The Democratic Congressional Campaign Committee, the D-Triple-C, made it a priority to put up a good fight in these five districts and called out the big guns. I was pissed off enough by the cutthroat and blatant gerrymandering that I violated my doctor's orders and my own common sense and agreed to work day-to-day on one of the campaigns.

I was rewarded for my stupidity with the toughest of them all: Tim Holden, a conservative Blue Dog Democrat who found himself living in the newly drawn Seventeenth Distict of Pennsylvania, which contained 66 percent of the district of Republican George Gekas and about 33 percent of Holden's old district. There were 50,000 more Republicans than Democrats in this district.

It was a sure loser. And I was completely fed up with broadcast politics. Still, Tim had a great team, guys like Bruce Andrews, who ran the campaign; Daren Berringer, the field director I would later coerce into being the state director of Michigan for Dean; and Tim Smith and Trish Reilly-Hudock, congressional staffers who did amazing constituent service work. Eventually, my outrage at the blatant Republican power grab trumped my reservations and I agreed to work on the Holden campaign. Actually, because I have no qualms in my own head about my sanity, I ended up begging to work on it.

By the time I arrived, it was already as ugly as anything I'd ever seen. As if they hadn't done enough by redrawing Holden's district, now they were coming after him with both barrels. We were being outspent three-to-one, and every morning we had three new attack ads to deal with. And since this damn district was only 140 miles or so from Washington, D.C., every day, it seemed, a helicopter would take off from the White House and deliver some special guest—President Bush, Vice President Cheney, Defense Secretary Rumsfeld, Secretary of State Colin Powell—to Pennsylvania to campaign for Gekas. All of this at a time when the country was firmly behind its war president and his campaign against terrorism.

When I get in a political fistfight like this, I don't pull any punches. I fight with whatever tools I have. I learned a long time ago that if you hold back, the good guy gets killed.

That's what was happening here. Tim Holden was a good guy, a good Congressman, and he didn't deserve to get pummeled like this. The adrenaline kicked in, and at some point, I just decided that Tim Holden wasn't going to lose. Not on my watch. So I moved to Harrisburg, Pennsylvania, and slept on the couch in the campaign headquarters or at the local Quality Inn. I walked precincts, rang doorbells, and worked myself ragged, just like the old days.

There were three partners in my media firm—Steve McMahon, Mark Squier, and myself—and we usually worked on different races at different times. While I was going through holy hell on the ground with Holden, I kept getting phone calls from McMahon, who was consulting with one of his longtime clients, the governor of Vermont, Howard Dean, who was considering a run for president.

"Hey," Steve said on the phone. "I think Dean's really serious about running."

"That's nice," I said. "Go to hell." And I hung up and went back to my trench. I'd been on this campaign for months, and still we were getting our heads bashed in. Every fall during election years, my blood sugar goes to hell, my nerve endings start to jangle, I get run down, and I get pneumonia. Like clockwork. And 2002 was no exception. I felt awful. Some mornings, I really wondered if I was going to get off the couch. I began a countdown when there were sixty days to go—sixty days and then I'm done. Sixty days and then I sleep for a year. Fifty-nine days and then I'm done. Fifty-nine days and then I sleep for a year. Fifty-eight days . . .

With thirty-eight days to go until election day, McMahon called again. "I really think Dean's gonna do it. You ought to get involved."

"Okay. Go screw yourself." I was in dying in a bunker, getting hammered all day, every day in the worst fight of my life, and McMahon was calling to ask me about doing a presidential? It was like calling a guy who's being bludgeoned to death to tell him you drove by a good cliff that he might want to jump off.

Eight days to go, McMahon called again.

"You want to know where I'll be in eight days, Steve? Sleeping. For a year. Now leave me the fuck alone."

As election day neared, the media began noting that the least likely of the five targeted Democrats to survive, Tim Holden, was actually making a move. In the last days, they were saying Holden was the one candidate of the targeted five with an outside chance of pulling off an upset.

Election night. With seven thousand votes counted, we were 15 percent behind. But all night, the vote kept creeping toward our side. Down to the last 10,000 votes, we were still behind. Finally, somewhere after 2 A.M. the morning after the election, we passed Gekas. It was one of the big stories of the mid-term—a miracle finish. Of the five Democrats targeted by redistricting, Holden—the one the numbers said was least likely to win—was the only one to pull it out.

Kathy and I had just bought a farm on the eastern shore in Maryland and I couldn't wait to actually *see it,* to spend some time there. Two days after the election, when I was getting ready to go to bed for a year, McMahon called again. He started by buttering me up, congratulating me on the amazing Holden win, telling me how it was the talk of Washington.

"Look," Steve said, "Howard Dean is out in Iowa. You gotta do me just this one favor. You *know* Iowa. Just take a trip out there, see what it feels like, hook him up with some people. Then you can go back home and sleep for a year. I understand. But before you go to sleep, you've really gotta hear this guy."

MY RESPONSIBILITY AS AN AMERICAN CITIZEN

Every cell in my body was telling me to say no. Don't do it! I had barely made it through the Holden campaign, the last thing I needed was a trip to Iowa. Then again, what could one trip hurt? (Addicts will tell you thats how it always starts.) So I flew out to Iowa, for an early campaign event in Des Moines. This was more than a year before the caucuses— and there were only two candidates there, my old boss Dick Gephardt and Howard Dean.

Gephardt went first and nailed his stump speech. And then Howard Dean got up and fell flat on his face. It was the worst speech I'd ever heard. He was an intriguing guy, sure. He seemed to be genuine, and unafraid. But honestly, I wasn't sure what the big deal was.

That night, we were driving back to Linn County for another event, and Dean told me what had happened. For weeks, he'd been out in Iowa by himself, campaigning with no other candidates around. But Gephardt had come out and given Dean's speech. Ironically, Dean was worried that it would look like he copied Gephardt's speech. So he made up a new one on the fly, went way off message, and came across looking discombobulated.

We talked in the car and I gave him a few ideas, how to move some things around, change the structure a little bit to differentiate himself from Gephardt. That same day, at the Linn County event, Dean got up first, and this time he blew the roof off the place. Gephardt and Kerry were both at this event, the first time all three had been in the same place at the same time. I looked over at Gephardt and Kerry, and it was just what I'd thought. They both had *Oh shit* looks on their faces. I knew what Gephardt was thinking: Where was that guy this morning?

In this environment where your patriotism could be questioned just for disagreeing, here was the former governor of a tiny state saying: *Wait a minute. We don't have the evidence to go to war.* I watched as Gephardt and Kerry sat there slack jawed, their mouths open. The looks on their faces said, *He's out of his mind.* Ninety-eight of ninety-nine senators had recently voted to cede civil rights to the Patriot Act—*with no debate.* The resolution giving President Bush authorization to attack Iraq had passed overwhelmingly. So Dean was courageous.

But still he didn't have me. Not yet. Having worked for Gephardt in Iowa, I knew he had a rule about dinners where there was more than one candidate: he would stay and shake hands and sign autographs until he was the last candidate there. He would always be the last one to leave.

So I hung around to see what would happen. The dishes hadn't been bused from the table before Kerry left. But Gephardt and Dean both stayed to work the room. They circled around each other, shaking hands, signing autographs, and talking—each occasionally glancing over to look at the other guy. Finally, there couldn't have been twenty people left and Gephardt gave up and went home. Only then did Dean leave.

So he was tough. I was coming around, but he didn't have me. Not yet.

The next day, there was another event, at a house in Linn County. There were probably 300 people there and I was looking around, thinking about old Iowa Caucuses, when someone asked Dean the oldest question in presidential politics: Why was Dean running? I'd heard the answer a million times: to get Americans back to work or to restore pride in America, to give you this and get you that. Whatever.

But then Howard Dean started talking. And this wasn't a prepared answer. I could see that he was genuinely answering the question.

"I'm running," he said, "because there used to be a time in this country when it wasn't enough that I wanted a good school for my kid or you wanted a good school for your kid. It was my responsibility as an American citizen and your responsibility as an American citizen that we all work together for good schools, not just for our kids, but for kids in Mississippi, Alabama, Oakland, California, and Harlem. I'm running because it's not just enough to want health care for my kid or health care for your kid. We have a responsibility as American citizens, a responsibility to each other, to provide health care for those kids from Oklahoma and Minnesota and Arizona . . ."

And that was the moment he got me. Every synapse in my body was yelling: *Get the hell out of there!* If you don't run to the nearest exit right now, you're going to start believing this guy. You're going to fall in love and by tomorrow morning you'll be working on this goddamn campaign.

The son of a bitch got to me. He talked from the heart about the things I believed. He said that America was drifting toward war in Iraq for the wrong reasons. That the country was too beholden to special interests. That the greatest nation in the history of the world ought to be able to provide health care for its people. I listened to this guy say the things that I'd been waiting to hear a candidate say for twenty years and like a damn schoolgirl, I fell in love.

I raced home and tried to forget about what I'd just seen.

I had a beer with my old friend Steve Murphy, who had just come off the Holden campaign, too, and we were commiserating about how tough the political consultant business had gotten. He told me that he'd heard I'd been looking into the Dean campaign. "You're not stupid enough to get sucked into that presidential shit this year?" he asked.

"No way. A presidential? I'm too old. No way I'm doing that. How about you?"

Steve shook his head. "No way. Never again."

Four days later I heard that Steve had agreed to run the Gephardt campaign.

Still, I held out. For weeks, I kept up the dance, taking trips to Vermont, advising the Dean people on the campaign, but trying to keep from committing myself. Each time I returned from Vermont, I was totally convinced: There was no way I could take this campaign. Dean was so far behind and his campaign was so unlikely, that to work on this thing day-to-day would be a death wish for me. Mike Ford even called me at one point and said that he'd heard I was thinking about working on Dean. "Don't do it, bro," he said. "I don't want you to be the next Tully."

I didn't need much convincing. No way was I in any condition to take on even the simplest presidential campaign, let alone a creaky, under-manned vessel like Dean for America.

And yet I found myself thinking about it: Maybe this *was* the guy. Maybe he was the real thing, the one you wait for your whole life. Maybe this was even the guy who could finally lead a real insurgency—a ground-up, honest-to-god revolution.

Stop it!

In February, after I'd been up to Vermont a couple of times as a consultant, Steve McMahon convinced me to go to Washington, where Dean was scheduled, along with the other six announced candidates, to address the Democratic National Committee's winter meeting, one of the first hurdles of a prospective presidential candidate and an important measure of his viability as a candidate. As befitting such a seat-of-the-pants campaign, he'd been scheduled all over the country in the days leading up to this event. It was the most important speech of his career to that point: a chance to define himself in front of the Democratic Party leaders, who would all be "super-delegates," an automatic four hundred voters at the Democratic convention in Boston. But his people had done very little in the way of promotion; while other candidates had brochures and pennants, Dean for America had virtually nothing on the tables or in the hall where he was speaking.

So while Kelly McMahon and Howard's small staff scrambled to put together a little Dean care package (the end result was perfect—a shot of Vermont maple syrup, Dr. Dean's Prescription-for-Change bottles filled with spare change—get it, Dean for *change*?) another small group huddled with the governor. They had been yo-yoing him around the country and the poor guy showed up with nothing prepared. Well, clearly this was not going to be some focus group-tested, TelePrompTer-delivered plea for party support. I guess that was good. But it had the potential of being a huge bomb. He and his aides were talking in hushed tones about how to deliver the usual Dean fare, health coverage for all Americans, when I spoke up.

"Look," I said. "You know what? I've been walking around, talking to these DNC members for a couple days. You know what they really want? They're waiting for someone to walk up to that podium and ask, 'What the fuck is going on here? What the fuck happened to our party?'"

It was not lost on Howard Dean that my first real advice to him consisted of a few lines of a public speech that couldn't actually be delivered in public. He looked at me calmly. "Joe, I can't go out there and ask, 'What the fuck is going on here.'" But I could also see he liked the idea. "How about if I say, 'What I want to know is—'"

And with that he took out a few index cards and made simple, one-word notes as we talked about how he should ask what the hell happened to the Democratic Party's principles over the last three years. He listened intently as Steve Grossman (the closest thing we had to a campaign chairman and former chairman of the Democratic National Committee), McMahon, Dean campaign manager Rick Ridder, and I offered our advice on the new speech. He nodded a few times, jotted down some key phrases on those index cards of his and put them away in his jacket. I thought: *That's it?*

Seven candidates were delivering speeches. When Howard's turn came, he looked around the room, stepped up to the microphone, and hit it out of the park. "What I want to know is why in the world the Democratic Party leadership is supporting the president's unilateral attack on Iraq!" The room erupted. "What I want to know is why are Democratic Party leaders supporting tax cuts . . . What I want to know is why we're fighting in Congress about the Patient's Bill of Rights when the Democratic Party ought to be standing up for health care for every single American man, woman, and child in this country. What I want to know is why our folks are voting for the president's No Child Left Behind bill that leaves every child behind, every teacher behind, every school board behind, and every property tax payer behind."

The room was liquefying. A staid, perfunctory election tradition had become a rally for the forgotten ideals of the *people's party*. Someone in the audience screamed out, "We want to know, too!"

And then—with a mixture of total joy and complete dread—I watched the ball sail over the goddamn fence. "I'm Howard Dean," he said in that forceful, matter-of-fact voice, "and I'm here to represent the democratic wing of the Democratic Party."[4]

The day was his now. The story of the winter meetings was Howard Dean, and the media took its first real look at him: at this candidate who had stepped up to tell the Party leaders that they had gone off-track and that he was calling them on it. In one twelve-minute-and-thirty-second burst, Howard Dean had announced that even if he might not win the race, he was going to make some noise. And I could feel myself slipping in behind him.

After that, the pressure picked up from every corner of my life. Steve McMahon had somehow convinced Kathy to fall for Dean and, just as importantly, she knew me, and she could see through my protests. So while she worried about my health, she also knew that I *needed* to do this.

I had even gotten a phone call from David Bender, the old campus organizer for Ted Kennedy, whom I hadn't talked to in twenty-three years.

---

[4] The first person to use this line was actually the legendary Minnesota Senator Paul Wellstone.

David said that he wanted to work for Dean, because as governor, Howard had signed the civil unions bill guaranteeing the most basic rights to gay couples in Vermont. "Oh, by the way, Joe," he said, "I'm gay." He told me that he'd met Dean several years earlier, when John Kennedy Jr., had brought the obscure Vermont governor to an editorial board meeting of *George* magazine, where David was a contributing editor.

David said that he'd read that I was thinking of consulting with Dean, and he wanted me to know he was ready to bring his considerable talents and contacts as a political and entertainment industry contacts to the campaign.

I told him that was swell, but I wasn't working the frickin' campaign. *What if this is the guy?*

It was an open secret that Dean and his top aide Kate O'Connor weren't happy with Rick Ridder, the capable, veteran campaign manager. People kept calling from Burlington to see if I was interested. I kept telling them no. I was in no condition to run a campaign. I didn't even know if I could do it. I didn't know if *anyone* could do it. (I found it ironic later when some people said that I hadn't effectively managed and organized the campaign. *Yeah, I told you that. Remember?*) I tried giving them other names, like Mike Ford, who I said could run circles around me. They'd be much better off with Ford. So Mike flew up to Vermont to take a look, but after a few days he turned right around and went back home. It was an impossible job, he said. Everyone was too green up there. The campaign was two years behind and falling further behind every day. No one was in charge, there was no organization, and the candidate didn't seem likely to do anything about it. It was suicide, taking a campaign like that. Not only wouldn't he take it but he said he wouldn't let me take it either.

But the pressure was coming from other places, from Steve and Kathy and David Bender, until finally, in March, the situation at Dean Headquarters came to a head. Word got out that if I flew to Burlington that night, the governor was prepared to ask Ridder to step aside and make me the campaign manager.

Every five minutes that day, my cell phone rang with people telling me to go to Vermont. They all said the same thing. I had to go right away. That night. Dean did this kind of thing on the spur of the moment and if I wasn't up there right now, the opportunity might be lost.

Kathy was standing at the door with my packed bag. "You know you want to do this. Go. Get on the plane."

I said no.

Outside the weather had deteriorated into a full blizzard. But that wasn't why I didn't want to go. I was still sick from Holden, and really did want to sleep for months. But that wasn't it either.

"Go ahead, darlin'," Kathy said. "Get on the plane."

I just stared at her.

Steve McMahon called. "Come on, Joe. Get on the plane."

"No," I said. "I'm not going."

Bender called. "You have to do it. You've got to get on that plane and you've got to do it tonight. If you don't do it tonight, it's not happening. Get on the plane."

I drove to the airport. But I couldn't board. I'd already missed one flight. I sat there, trying to figure out how to get out of going up to Vermont and taking this job. There was only one more flight that night when I called David back.

"Look," I said. "This is gonna sound crazy, but I need to tell you something and I might need you to talk me through it. Ever since I was a kid, I've had this horrible nightmare.

"In the nightmare everyone I know is telling me to get on a plane. It's really important that I get on a plane. So I do and the plane takes off. And it crashes. I know it sounds crazy, David, but this feels like that moment."

Now I'm a scientific guy, an empirical thinker. I don't get spooked easily. But that night, I was frozen with fear. I was sure my plane was going down. Maybe it was my subconscious trying to tell me something, that with my precarious health, a presidential campaign could kill me. Then again, maybe the plane really was going to go down.

David said that I wasn't crazy (although, remember, he hadn't seen me in twenty-three years) and that if I really felt that way I should probably turn around and go home. No one would blame me. The world would go on.

What if this is the guy?

The truth is that there are only a few people who ever get the chance to run a presidential campaign, just like there are only a few people who get to play shortstop in the major leagues. I love baseball, but that was never *my* dream.

There was a slight chance, however, that my dream was sitting up in an office in Burlington, Vermont. And I might never know unless I got on that plane.

"Screw it," I said to David. "I never let that nightmare get to me before. I'm forty-six years old. I'm sure as hell not going to let it get to me now."

I walked toward the jetway. The journey I was about to begin would be the most amazing and unforgettable of my life.

I would be witness to the beginning of a tiny movement that would quickly grow to hundreds of thousands of Americans, the first step toward rebuilding the most fundamental cornerstone of this country, one that had been lost in the slick cynicism of broadcast politics—the profound and absolutely vital involvement of the people in choosing their own government.

The Dean plane would eventually go down . . . a year later in an Iowa cornfield. But by then, it wouldn't matter. The revolution would be well underway.

PART II

THE PLACE WHERE
THE FUTURE HAPPENS

5

VERMONT
Phish, Chicken Dinners, and the Deanie Babies

URLINGTON, VERMONT, IS a beautiful, bucolic town of 39,000 gentle, fleece-wearing souls, the rustic birthplace of Ben & Jerry's first ice cream store and the counterculture jam band Phish. It is also—without a doubt—the stupidest place in the world to launch a revolution.

Vermont is the second smallest state in the union, home to just more than 600,000 people—about the same number as live in El Paso, Texas. The capital of Vermont, Montpelier, has only 8,000 of those people; you could fit them all in an arena for a small Phish concert. With the possible exception of when one of its senators switches political parties, Vermont exerts no real political gravity on the rest of the United States. On the rare occasion the state is mentioned at all by the national political press, it's invariably dubbed "the People's Republic of Vermont," an eccentric, liberal, out-of-the-mainstream, bed-and-breakfast commune—Berkeley in a ski sweater. No election-year pork or vice presidential candidate has ever been floated with the idea of wooing Vermont voters. No electoral strategy has ever begun with the Green Mountain State and no presidential race has ever dramatically turned there. (Among the things you'll never hear in a campaign: *We were doing great until Vermont.*) Its people take great pride in being politically independent, progressive, and unpredictable—an island of rugged northeast individuality. This makes Vermont a great place to live, but the last place you'd think to go to begin a movement, since it lacks the first thing you require for a populist insurgency: a populace.

Like the state he governed longer than anyone else, Howard Dean doesn't seem, on first glance, like an obvious candidate for a grassroots, power-to-the-people insurgent campaign for president. Or on second glance, for that matter.

Born to privilege on Manhattan's Upper East Side, Howard Brush Dean III was just another twenty-four-year-old Yale graduate on Wall Street, on his way to becoming the fourth generation of Deans to make his fortune as an investment banker, when he suddenly changed course. Uninspired by the thought of spending his life moving money around and driven by a deep desire to help people, he left Wall Street in 1973 and began secretly volunteering at a Greenwich Village hospital. It says something about the expectations in the Dean clan that his resulting decision to go to medical school was one he had to *break* to his family.

Howard was taking pre-med classes at Columbia University when, in the summer of 1974, his brother, Charlie—an outgoing volunteer for George McGovern's presidential campaign two years earlier—was captured and eventually killed by Laotian guerillas while traveling with a friend through Southeast Asia. Charlie's death affected Howard deeply and permanently (he wore Charlie's belt every day of the campaign) and underscored his desire to live a life of meaning and service. He graduated medical school in just three years and moved to Burlington, Vermont, in 1978 to begin his residency at the University of Vermont. Three years later, he married Dr. Judith Steinberg, whom he'd met in medical school. They set up a family practice in Burlington.

He got into politics right away, over the quintessentially Vermont issue of whether a hillside overlooking Lake Champlain would be better served with a bicycle trail or by some condos. He smartly went with the bikes and was elected to the Vermont state legislature in 1982. In 1986, he won his first of two terms as lieutenant governor, a mostly ceremonial job that rarely interfered with his medical practice. When the governor, Richard Snelling, died of a heart attack while cleaning his swimming pool in 1991, Dr. Dean stepped in admirably as governor, erased a $65 million debt (this in a state whose total budget is only $1 billion), and quickly established an amazing record as a disciplined, bipartisan, nononsense leader, frustrating ideologues on both sides with his logical, confident physician's approach to governing. By avoiding either side's politicized rhetoric, he was able to pass some of the most progressive laws in the country by selling them as commonsense measures, and in the process made Vermont an example for the rest of the country—extending the most basic rights to gay and Lesbian couples, providing a more equitable system of funding public schools, and extending health care to 97 percent of Vermont's children and 91 percent of its adults.

Over four more terms, he grew even more popular and more respected, not as the oratorical spark of a populist *movement* (the thing he would improbably become), but as a pragmatic, get-the-job-done regular guy. In fact, in my opinion he became one of the greatest governors in the country's history, and as such, was never seriously challenged in an election, and never had to spend more than a million bucks getting reelected. This was great for Governor Dean in the 1990s, but not so good for *Candidate Dean* in 2003, who had never run a serious state race—the equivalent of a 200-meter dash—let alone the marathon of a presidential campaign. And so he arrived in a national election with no national plan, no national team, no money, and next to no campaign experience—seriously, there were freshman members of Congress who had more tough races under their belts. And no help in sight. No one really thought he had a chance. In a story in *New York* magazine, one observer described the Dean campaign as "quixotic," "preposterous," and "the silliest thing I'd ever heard." That observer was Howard's own mother.

In those early, empty-wallet days of the Dean for America campaign, anyone coming to Burlington from Washington, D.C.—like I did in the winter of 2003—felt obliged to pass on the $580 flight to Burlington in favor of Southwest Airlines' discount $79 ticket to nearby Manchester, New Hampshire.

That meant borrowing a car and driving three hours, across half of New Hampshire, over the Connecticut River into the white pine and sugar maple forests of Vermont, winding up Interstate 89 past White River Junction, Bethel, Montpelier, and Waterbury, between the rolling hills of the Winooski River Valley, until finally Burlington appeared in the trees like the last civilized outpost before the wilds of Lake Champlain and the Adirondack mountain range.

When you arrive in Vermont the first time, everyone says the exact same thing to you: "Welcome to Vermont, where we have eight months of great skiing . . . and four months of really lousy sledding."

And the thing is . . . they mean it. In 2003, for instance, it didn't stop snowing there until May 29, and immediately started up again on October 1. I arrived that winter to find everything blanketed in a sleepy cover of snow, especially Howard Dean's presidential campaign.

That January (two months before I would take over as campaign manager), the Dean campaign was still squirreled away in a cramped, 1,000-square-foot second-story office above the dark Vermont Pub and Brewery. There were six people—seven if you counted the governor—working for Dean for America, most of whom had been longtime aides in the governor's office. They were, for the most part, smart, energetic, good people, committed to Dean, but their experience in politics came from governing within a 400-mile radius, in a state where the biggest pressure lay in keeping the governor prepared and on time for a town hall meeting in Montpelier. By January 2003, one year before the Iowa primary, while the other campaigns had built sophisticated political machines, raised war chests of millions of dollars, and compiled computerized lists of potential supporters in key states, the Dean campaign had none of these things, had raised only $315,000, and had spent two-thirds of it just remaining on life support.

Dean for America, in those early days, was euphemistically described by one writer as "charmingly modest." Or, put another way, broke.

The day I climbed the stairs past the Vermont Pub and Brewery I couldn't believe this was a presidential campaign one year before the Iowa caucuses. There was a computer in the Dean headquarters—and a relative of the governor's had set up an early web site—but it wasn't even turned on. They had gathered about 9,000 names of "Friends of Howard," people who had, at one time or another, told the governor that they might be interested in helping if he ever decided to seek higher office (although few of those people would have guessed that he would go *this* high). But instead of being readily accessible for sorting on a computer database, these names, along with names of thousands of other potential supporters, were scrawled on business cards, contact sheets, and scraps of paper and stuffed in a few shoeboxes—not even one shoebox for each state.

In his book about the early part of the campaign, *One Car Caravan,* *USA Today* political columnist Walter Shapiro titled his first chapter "Rubes on the Road" and opened with a description of Dean's "seductive fantasy" of being the Democratic nominee, "outlandish as it seems for a candidate without money, campaign staff or national following."

In fact, during Howard's first months as a candidate, he essentially served as his own campaign manager and political strategist (bringing to mind the old saying about the man who serves as his own lawyer . . .) and the press spokesman was whoever happened to answer the phone. His only traveling aide was his rail-thin, mop-haired, fiercely loyal thirty-nine-year-old chief of staff Kate O'Connor, who had flitted around Dean as his closest aide since he was the part-time lieutenant governor in 1989, and who sometimes acted as if she was on summer vacation—mailing post-cards back from Iowa ports of call and filming the whole thing on her personal video camera.

Throughout 2002, while the other campaigns were slow-dancing with the top political operatives in the country, hoping to secure their services for the 2004 run, Dean held back, trying to conserve money and assuming the top guns wouldn't come sign on with such a long shot anyway. Finally, he did pick up a few aces, including the very able Rick Ridder, who had worked for Gary Hart and Bill Bradley's presidential campaigns, and whose strategy was the sound, time-tested idea that Dean begin by courting the traditional Democratic leaders and bases (or by courting someone . . . *anyone*). He also got Steve Grossman, the former national Democratic Party chairman who hailed from Massachusetts and threw away years of close personal friendship with Senator John Kerry the day he signed on with Howard Who? And Stephanie Schriock, his amazing finance director—young, smart, tough as nails, and the only person in the campaign who could scare the daylights out of me when I made her mad. When Grossman signed on the entire political establishment thought he had lost his mind. When Schriock told her friends she was heading up to Burlington to work for Dean, they told her it would be the end of her career.

Dean was getting very little media coverage in those cold, dark Vermont days, other than those stories about how little chance he had. The first Dean stories all had headlines like "The Invisible Man" and "The Darkest Horse."

That January, Dean boldly told reporters that he intended to raise a whopping $10 million, a fraction of what the frontrunners would raise. This number—small by presidential campaign standards—still managed to sound like a pipe dream coming from the tiny Dean office (it was like the painfully behind-the-times Dr. Evil in the *Austin Powers* movies threatening to blow up the world unless he is paid *"one **million** dollars"*). While other candidates moved around on private jets, Governor Dean carried his own luggage on Jet Blue and Southwest flights and, rather than pay to stay in hotels, he spent nights sleeping in the spare rooms of his supporters' houses.

In most polls, his "support" was less than the margin of error of the poll: 2 percent here, 1 percent there. When I arrived in January, Dean had been campaigning in Iowa by himself for months, and yet he was tied there with the Rev. Al Sharpton at 2 percent, badly trailing the "serious candidates": Gephardt, Lieberman, Kerry, and Edwards, who hadn't even fired up their massive campaign machines yet. Most distressing, after months out there alone, he had by far the lowest name recognition—20 points worse than any other candidate. Forget supporting him, in January 2003, 82 percent of Iowa voters couldn't say who Howard Dean was.[1] Even Iowa Senator Tom Harkin had a problem identifying him back then. At a forum he had painfully introduced the governor as "John Dean."

I had been brought in to assess how the Dean campaign was progressing. But looking at the organization that January, at the message, at the polling numbers, and, especially, at the fund-raising, I couldn't help wondering when the damn thing was going to begin.

Even for the type of campaign Dean wanted to run—a classic insurgent riding in from outside the political establishment (my specialty)—he was two years behind where he needed to be. And there was only a year left to go before Iowa. Some of the other candidates had been preparing for this run—raising money and gathering supporters—for years. I knew for a fact that John Kerry had been considering a run in 2004 for *fifteen years*. I'd sat with his close adviser Ron Rosenblith in 1989, and brainstormed ways in which he could build up Kerry's direct mailing list. I listened as Ron explained that if Kerry didn't run in 1992, 1996, or 2000, that would almost be better, because the list would be that much longer in 2004. By the winter of 2003, these campaigns were armies, with a hundred staffers, computerized databases, and millions of dollars at the ready.

We were seven people sitting around Vermont with a bunch of shoeboxes.

And yet, even with its stunted beginning, there was undeniably something appealing about Howard Dean's quaint operation—something *real*.

---

[1] Zogby International Poll.

The resolve of Howard's team, the candidate's refreshing honesty and lack of political guile, and the sameness of the other candidates all gave Dean the whiff of a true insurgent. The challenge was finding some way to fast-forward the usual campaign building and, at the same time, skip over the dismissive TV media and appeal directly to the American people.

Talking about the Democratic field at the time, James Carville, Bill Clinton's old political consultant, mentioned Dean as an intriguing after-thought: "If he could raise money, he'd be dangerous."

SYSTEM FOR SALE

There is a dirty little secret in the world of auto racing: the races are often over long before the cars get on the track. The team with the most money just builds the fastest car and gets the best crew, and it's over the minute they unload the car from the trailer. That driving around the track business is mostly for show; the race was all but over when the engine was built. When you've got the best car, only driver error, serious tactical goof, or an act-of-God crash can lose the race for you.

This is true of politics as well (substitute sex scandal for car crash). Since 1960, elections have become a race to see which team can get the most wealthy people and corporations to donate the most money to their candidate. The team that raises the most money buys the most television ads, then uses these blunt instruments to pummel the most Americans into *not voting* for the other candidate. And, as long as the driver doesn't screw it up, the guy with the moneyed car is the one who cruises around the track to victory. Between 1976 and 2000, the Democratic and Republican presidential candidate who raised the most money and qualified for matching federal funds in the year before the primary season, was the party's nominee *every time*.

In the book, *The Buying of the President 2004*, Charles Lewis, the executive director of The Center for Public Integrity, a nonprofit, non-partisan research organization, writes that the focus on early fund-raising ultimately takes the democratic process out of the voters' hands, that "the race for the White House is substantially decided before any votes are actually cast. The dirty little secret of American presidential politics is that the wealthiest interests essentially hold a private referendum the year before the election."

As George W. Bush has single-handedly proven, when you get a few hundred people from Enron to donate two grand each, there's a good chance the mother ship might want something for its half million bucks once the race is over. When added to the massive corporate donations made to special interest groups (whose campaign spots often do the candidate's dirty work for him), the result is a system in which both political parties, and most candidates, are for sale. Bill Brock, former chairman of the Republican National Committee, may have put it best when he testified under oath in a lawsuit over campaign finance reform, that:

> These contributions compromise our elected officials. When elected officials solicit these contributions from interests who almost always have matters pending before the Congress (they) become at least psychologically beholden to those who contribute. It is inevitable and unavoidable. The contributors, for their part, feel they have a "call" on these officials. Corporations, unions and wealthy individuals give these large amounts of money to political parties so they can improve their access to and influence over elected party members. Elected officials who raise soft money know this. The appearance of corruption is corrosive and is undermining our democracy.[2]

This is how we've ended up with such blatant transactional politics—fund-raising as pure transaction between the candidate and the moneyed interests who navigate campaign laws to float his candidacy, with the full expectation of a seat at the governing table (or in the case of the Bush administration: the table itself, and the chairs, and the rugs . . .). This system in turn empowers the corporate and special interests, who provide another source of the fuel these campaigns are really after: cash. Each year, more money is needed for TV ads—an estimated $1.6 billion this election cycle—and the new fuel simply makes the thing that much more corrupt. This is how we ended up in 2004 with thirty-three lobbyists for every member of Congress, a presidential election with a totally detached electorate, campaigns run exclusively through TV ads, and filthy rich candidates beholden to special interest groups.

---

[2] Charles Lewis and the Center for Public Integrity, *The Buying of the President 2004* (New York: HarperCollins, 2004).

"If you look at the list of candidates now who are prominently mentioned for president, almost all of them who have any chance at all are
millionaires or multimillionaires," former President Jimmy Carter said
recently. "An average person like I was, just a peanut farmer back in
1976 (running for president)—would be absolutely impossible, which
means that there's a criterion for success in American politics now—the
Democratic or Republican Party—and that is extreme wealth or access
to major wealth. And we are the only democratic nation in the world, in
the western world, within which that blight or cancer is affecting our
system."[3]

There are few things I hate more than this self-defeating system of
politics, and I am confident that one day soon it will be lying in a heap
alongside the road (along with other artifacts from the television age), but
in the winter of 2003, it was the only game in town. So my challenge as a
consultant—and later as the campaign manager—was to figure out how to
get Howard Dean's car on the track, how to raise money, raise our candidate's profile, and create some momentum going into Iowa . . . and how
to do all this with no time, almost no support, and no framework in place
for a national candidacy.

Early on, we had a meeting to figure out how we were going to bring
the Dean campaign around. We went through this exercise in which we
identified eight basic strategies, wrote them on a white board, and went
through what it would take to accomplish each of them. It didn't look good.
George W. Bush was in the process of raising some $200 million for his re-
election, so we figured that if we somehow won the Democratic nomination (putting aside for a moment the notion of hell freezing over) we'd need
our own $200 million to make a serious run at the presidency. We could
look for help from political action committees of course, but that was the
antithesis of the Dean campaign—like asking the oil companies for money
to create alternative fuels.

The other campaigns were going the $2,000 chicken dinner route, but
we knew there simply weren't enough rich people for us to hit up for the
money. So maybe we could appeal to regular Americans, get two million of
them to donate a hundred dollars each. If we could have a dinner each night
where one thousand real Americans donated a hundred dollars each, we

---

[3] Ibid.

could raise a million dollars in ten nights. In a hundred nights, we could raise ten million. In a thousand nights, we could raise $100 million. In two thousand nights, we could raise $200 million to defeat George W. Bush.

Of course, two thousand nights is five-and-a-half years and George W. Bush would be retired on his ranch.

So right away we could all see that our only hope for winning now was to decentralize the campaign, ease control away from the candidate and his handlers in Vermont (myself included), and let the momentum and the decision making come from the people—stop trying to control the river . . . just open the flood gates and see where the current took us. Howard Dean recognized this himself—after every trip he would come off the road and tell me we had to find a way to decentralize the campaign (his word, not mine). Decentralizing was the only chance he thought we had. We would never have enough money to build the campaign right, and the more he kept asking me to find a way to decentralize things the more I got excited.

Like someone whose entire life has been building to this point, I knew without looking what our only hope would be: the Internet. I had been waiting for this moment for a long time. In a story by Noam Scheiber in the *New Republic,* the veteran Democratic campaign strategist Bob Beckel recalled the first time he heard me talking about it.

> Beckel remembers doing a panel discussion with him not long after the 1984 campaign, when Trippi was already talking about an early version of the Internet and how it could change politics. "I said, 'Joe, I don't have any idea what you're talking about,'" Beckel recalls.

Now, almost twenty years later, I still didn't know if the Net would be up to the task. When John McCain had tried to use it to drive his own insurgent campaign in 2000, the technology had only proven mature enough to create a small buzz around an interesting candidate. Not enough snow had been plowed by Amazon.com, eBay, and all the online travel agencies for a political candidacy to make much headway. Making matters worse, in the four years since McCain's short run, the Internet bubble had burst—tech stocks losing more than half their value in the stock market—and with it had gone much of the futuristic optimism that had once powered the Net as surely as electricity and computer chips.

But what other tool did we have to take our campaign directly to the people in such a short amount of time, to build the campaign virally—one person infecting two more, who infect two more and two more—to let that water rush in and see if our boat was ready to float.

So while I was excited to finally be launching my first grassroots, Internet insurgency, honestly I didn't give it much chance of success. Best case scenario: We'd make a little noise as a quirky sidelight to the campaign, quickly fade away, and in three months I'd be back home on my farm.

And so, on my very first day in the Dean campaign headquarters, that January, I offered up the closest thing I had to a strategy: "We need to put a link to this web site, Meetup.com, on our campaign web site."

I had come across the fledgling Meetup by accident, when I was trolling around Internet web sites and blogs. One night, months earlier, I had visited a blog called MyDD.com and read a posting by a guy named Jerome Armstrong, who was commenting on the early presidential campaign season and specifically, an idiotic quote he'd read from some know-it-all political hack: me.

I fired back, "Hey Jerome. It's Trippi . . ." and defended myself. Pretty soon I was reading Jerome's blog regularly, and occasionally commenting on my own stupidity. Then, in January, right before I went up to Burlington, Jerome wrote in his blog about a web site called Meetup.com where, he said, some Howard Dean supporters were using the Internet to get together in a handful of cities.

I jumped over to the Meetup site.

Despite its perfectly descriptive name (it helps people—*meet up*) every time I tried to explain the concept to someone in early 2003, I would get that blank stare, the same look I got twenty-two years earlier when I tried to explain to Tom Bradley's people that it might help us with organization to buy one of these newfangled computers, or when I first brought up political organizing on the Internet to Bob Beckel.

They just didn't get it.

Meetup.com is simply a web site where people of similar interests (it could be anything: stamps, Star Trek, Howard Dean) are matched and given a time and place to meet, organized by the web site, which reserves a public place in each city (Starbucks are very popular) and notified all of the members of that interest group of the time and place.

As small as this thing was (most meetings were just a handful of people), it was exactly the democratic vision of the Internet that I had always believed in, using this technology as a way for people of similar interests, passions, and causes to find each other and *instantly* form into communities—tiny little Iowa caucuses made up of science fiction fans and curling enthusiasts and knitters.

Meetup.com had only been in business for a short time when Jerome brought it to my attention. But people were already using it to organize for the various candidates. And the first thing I noticed was that Howard Dean—dead last among the Democratic candidates in almost every other meaningful measurement—was actually leading in this one category, the number of his supporters who wanted to meet up.

It was a tiny number, 432 supporters across the entire country— maybe one in Seattle, three in Los Angeles, a couple in New York—but compared to the other candidates, he was killing. There were only 310 people who wanted to meet up for Kerry, 141 for Edwards. The only guy who didn't have an area code number of supporters was Dick Gephardt, who had something like 40 people signed up to meet and talk about him.

I stared at these numbers and wasn't entirely sure what to make of them. Clearly, this wasn't a representative sampling. At the time, Kerry's nomination was assumed to be a done deal, sewn up, with Gephardt second. And yet, in this one arbitrary measure, Howard Dean—a guy with maybe two percent of the vote, a guy 95 percent of Americans couldn't pick out of a criminal lineup—was drawing the most interest.

Then, an amazing thing happened. After we put Meetup on the web site, I checked back, and suddenly there were 2,700 people who wanted to meet up for Dean. The number had taken one of those exponential leaps—what would turn out to be the first of many. The second-highest candidate, Kerry, had only gone up to 330 names. This was more than just a statistical quirk. Something was going on out there. About that time, the visionary developers of Meetup.com contacted me and we negotiated an agreement for them to continue organizing the governor's supporters for a fee of $2,500—not a bad initial investment for a site that would eventually boast 190,000 Dean members. Later, as the other campaigns caught on to the phenomenon, they would call the Meetup guys and ask for what had come to be known as the "Trippi Special." The other campaigns were about six months too late. No one was ever going to get that deal again. On the Internet, being the first mover can be everything.

People often ask me which came first: the Dean campaign's embrace of the Internet or the Internet's embrace of Howard Dean. The answer is that it was a little bit of both. The curtain was rising on the Internet political movement right about the time we made the decision to turn on the lights.

But we almost missed our cue.

A couple of days after I had suggested putting Meetup on our candidate's web site, I checked and the link still wasn't there. The resident computer expert (and the only one), a great guy named Bobby Clark, explained that some people in the campaign didn't think it was a good idea to put up a link to another web site.

"That's what the Internet is," I said, "a bunch of web sites linking together. Just put it up there."

It's easy to imagine the Dean campaign as always being plugged in, but in the beginning, nothing could be further from the truth. The candidate himself was a self-described "technophobe" who didn't have cable TV, didn't like to use a cell phone, and had only been using e-mail since 2001. Early on, he was the only candidate who didn't travel with a computer. The first time I talked to him about having a campaign blog, he admitted that he didn't know what that was. And so, early on, the campaign reflected its candidate's gun-shyness about computers and those nine thousand names remained on those scraps of paper in shoeboxes somewhere under Kate O'Connor's desk. Kate and I shared an office throughout the campaign—where she made me feel welcome in those early days by marking a line with masking tape across the floor like the equator that I was never supposed to cross.

Getting the campaign to use its one untapped resource, the Internet, proved to be harder than I thought. I talked myself hoarse explaining my idea for decentralizing, turning the campaign over to the people out there, giving them the tools and the support, and letting them do the work. But early on, the only person I could seem to convince was the in-house computer whiz, Bobby Clark. And he was the one person already on my side.

Several days after Bobby and I began suggesting it, Meetup still wasn't on the Dean web site.

"What the hell are they thinking, Bobby?"

Apparently, the campaign brass had decided to run it past the lawyers, just to make sure there were no legal issues having to do with in-kind contributions.

Lawyers? In-kind contributions? Christ. Here we were, running a seven-person campaign in frickin' Vermont, hugging the margin of error in the polls, and we're going to play bureaucracy? We're dying here and we want to spend our last hours talking to lawyers?

Go home, I told myself. Go home and sleep for a year. I was just here for a couple of weeks, to work with the scheduler and the political people, to give them some tips on running a presidential. I didn't need this.

Finally, after a week of my pestering, the link to Meetup.com went up on the web site. Now people who logged on to the Dean for America web site were given the opportunity to go to a place where they could actually get involved in the campaign and—even more important—find other people to get involved with them.

Within weeks, the number of Meetup people had shot up to 2,700, then 8,000. And this burst didn't come from the campaign buying a TV spot or scheduling speeches—in fact, this wasn't the campaign at all. This was the people taking over. For months, Meetup.com would run its own parallel campaign, the number of people meeting up growing from that initial 432 to more than 190,000. Eventually, we'd even have to create our own, specialized version of Meetup—the GetLocal tools, which would grow to 170,000 people on its own.

So how did we drag our feet and almost miss what would become the most effective organizing tool of the Dean campaign?

The same way the other campaigns, companies, and corporations missed it and continue to miss it. Forty years of reliance on television advertising has atrophied creativity, forcing everyone to approach every problem the same way. In politics, for instance, television advertising is seen as the only solution to every problem, and so far more thought is put into the menu of those $2,000 fund-raising chicken dinners than in how to actually get people *involved* in the campaign.

It may sound like just another new age, techno-buzzword, but the key really is *empowerment*. The fallacy of polling is that one person always represents one vote. Anyone who has seen an Iowa caucus up close knows that is not true—knows that there are Iowans, and then there is someone like Jimmy Hogan. Those 432 Dean supporters were far different than the people we would reach by television. These were the Jimmy Hogans, the organizers, and connecters, 432 individual campaign managers who got on the Internet and began doing what I could never do.

THE CHILDREN'S CRUSADE

They just started showing up. Kids mostly, in parkas and sundresses, in cargo shorts and hooded sweatshirts, clean cut and shaggy, some with piercings or odd bits of facial hair, others with briefcases and resumes, some with a bit of political experience, others with experience at Frisbee and video games. As soon as our campaign went up on the Internet, we discovered a small but intensely devoted group of people there waiting for us. And now they began to take the campaign into their own hands, organizing and gathering other people. And a few of them even got in their cars and began showing up on our doorstep.

That spring and early summer, young people from all over the country took a break from college or quit their jobs, hoisted their backpacks, and drove to Vermont to work on the Dean for America campaign. Someone aptly named them *Deanie Babies*. I had never seen anything like it; politics hadn't seen anything like it since 1968, the days of Bobby Kennedy and Eugene McCarthy's "Children's Crusade."

In Vestavia Hill, Alabama, an earnest nineteen-year-old named Gray Brooks listened to Howard Dean once on the radio, got on the Internet to research him, decided Dean was "a good man," and the next thing you knew, was in his car and on the freeway, driving to Vermont to volunteer for the campaign, to work seventeen-hour days and sleep on the floor. A handsome, clean cut, lifelong Boy Scout and Baptist, Gray couldn't help but be courtly, calling everyone ma'am or sir. Even me.

"Look," I'd say, "I swear to God, if you call me sir one more time, I'm going to fire you. Got it?"

And he'd say, without irony, "Sorry, sir."

I would think of Gray often later in the campaign, when Republican operatives, our Democratic opponents, and some in the press caricatured young Dean supporters as pierced, vegan weirdos. Here for the first time in decades were young voters inspired to get involved—and believing that they could make a difference—and they were being painted as some kind of freak show. For wanting to be involved! It was like the baby boomers in power had forgotten how long their hair used to be. Like they believed they were the only ones who could try to change the world. Here, finally, was another generation rising up, getting involved in politics, trying to stop an unjust war and take back a corrupt system. If anyone should have embraced this new era of activism, you'd think it would be the former children of the 1960s. Instead, they ridiculed these young people. This pissed me off more than anything else in this campaign. Didn't these people learn a frickin' thing? Didn't they watch their own movie?

All spring and summer, young people like Gray were drifting in from across the country. They were undoubtedly drawn to Howard Dean by his lonely opposition to the war in Iraq (as young voters of my generation flocked to Kennedy, McCarthy, and McGovern) and his call for a new kind of politics. But they were also drawn to the Dean campaign because someone was finally taking the time to reach out to them where they lived. Studies had just begun to show that young people were spending as much time on the Internet as watching television—and more in some cases. This represents a profound shift in what America is and, even more to the point, what it's becoming. As one of the first mainstream organizations to tap into this demographic vein, we were learning that, unlike generations of dulled and deadened TV watchers, these young people actually *wanted* to be engaged politically. They were out there asking questions, organizing, and just waiting for someone who could speak their language back to them, the language of the Net.

Not all of them came to Vermont to get involved, of course. Students for Dean groups, with no connection at all to our campaign, began popping up on college campuses, thirty by March, one hundred eighty by July. And not all of them were young. Middle-aged people, elderly people—anyone with a computer was able to join the discussion and once you joined the discussion, you had effectively joined the campaign, because eventually, *the discussion was the campaign*. The campaign was what the bloggers helped make it. We may have grown to a staff of thirty or so by March, but there was no way a staff of thirty could match the brainpower of 22,000 engaged Americans, all sharing ideas and urging others to join the cause.

Soon we had people volunteering to work on the ground in their own communities, and by spring, a campaign that had no national structure had volunteers stepping up in all fifty states—most of them attracted not by television—the old flaccid warhorse of political campaigns—but via the sleek, hungry Internet.

One day, soon after we'd moved to a larger quarters in a South Burlington office park, I looked up to see this tall young guy with an earring and a nearly shaved head wandering around the office. Security had just grabbed him and was hauling him away when he yelled out to me: "Wait! I blog on MyDD.com!" This was, of course, the political web site where I'd first heard about Meetup.com.

"You're hired!" I yelled, and they brought him back in to me.

It turned out his name was Mat Gross. He was thirty-one, married, from Moab, Utah, and he'd been calling Dean headquarters for months, trying to get a job with the campaign. Finally, he'd just decided to come to Vermont. He hadn't even packed, he just hopped on a plane. He'd hung around for a couple of weeks, trying to ingratiate himself, but hadn't had much luck. I hired him on the spot to be content director for the Dean for America web site, on one condition: before he went back to Utah to get his things he had to create a blog (for the uninitiated, this is short for "web log"—the increasingly popular, online daily journals and amateur reporting about anything under the sun, in this case, about a presidential campaign). I told him that as soon as he got a campaign blog up and running he could go back to Utah, get his things, and come back to work.

Within 48 hours he had created "Call to Action," the first-ever blog of any presidential campaign. I could see two things right away: (1) Mat knew what he was doing and (2) he must really have been in a hurry to get back to Utah to pack. This was the ugliest, messiest, unfriendliest site you've ever seen. There wasn't even a place for readers to add their own comments. It was like an AMC Pacer when compared with the sophisticated, interactive BMW blogs that we'd be driving a few months down the road.

And yet, even later, when we were in the process of designing its fully realized sequel, "Blog for America," we came to the shocking realization that we'd all grown strangely attached to it. It had been there from the beginning, loyal and true, like that twenty-year-old gimpy, blind dog that you keep around even though you know the most humane thing would be to just put it down. It took us months to finally euthanize Call to Action.

At first, Rick Ridder and I had worked side-by-side, but by March I had taken over the reins as campaign manager and I immediately set out to beef up the Internet side of the campaign. When I had arrived, there was one person devoted to the entire category of "computers"—Bobby Clark. I quickly picked off one of the most talented people on the campaign, a dynamic young death penalty lawyer and Vermont native named Zephyr Teachout, who became the director of Internet organizing and one of the campaign's moving forces. But I was always on the lookout for more help, and now, when these young people would straggle in from the road, my first question to them was whether they had any experience with web sites, blogs, or e-mails.

Along with Zephyr and Mat, the most important hire of that time was a guy named Nicco Mele, a caustically funny and bright graduate of William and Mary College, whose foreign-service parents had raised him in Ghana and Malaysia. David Bender had worked with Nicco at Common Cause and called to tell me that he had the perfect candidate for the still-vacant Dean webmaster position. Despite David's somewhat questionable track record for spotting talent (he'd hired me), I hired Nicco on the spot, and he didn't disappoint—hopping in his Honda and driving straight up to Burlington. He turned out to be the perfect blend of old and new politics: a big, bright, tireless, larger-than-life personality with an amazing knack for reaching people on the Internet, and a natural understanding of an insurgent presidential campaign. If Paul Tully or Mike Ford had been born twenty years later and gotten into computers, they might've ended up being Nicco Mele.

One of my favorite bloggers in the early days of the campaign was the "CarlwithaK" blog—written with wit by none other than Karl Frisch from his Dean outpost somewhere in California. One day I got on my laptop and began to make daily rounds to my favorite blogs and when I got to CarlwithaK I couldn't believe the post I was reading—something like "hey everyone, I have been blogging for a week now from Dean HQ in Burlington, the place is amazing . . ." I was stunned. CarlwithaK *was in the house* for a frickin' week and I had to read about it on his frickin' blog? "Get his ass in here," I said. "I want to talk to him." Karl would turn out to be another great addition.

With these singular personalities and the constant influx of new young people, the web room buzzed with life, the heart of campaign headquarters. Frisbees and Nerf balls flew around the room; there were squirt gun fights; skateboards and scooters flew down the hallways, all of it under the watchful eye of Nicco, who loomed over the room with his easygoing, natural leadership and a dazzling collection of yoyos.

Their energy worked on me like a drug. I loved imagining the quiet, professional offices of the other campaigns, twice as big and half as interesting. It was as if I were on my first campaign again—driven by these kids' idealism and by the feeling that we were changing the world every time we came to work. I contributed slogans from my own ancient youth, like the Gil Scott-Heron anthem *The Revolution Will Not Be Televised,* which quickly became the motto of the web room, written on screen savers and signs posted around the room. I told the Deanie Babies about the 1969 New York Mets (with the harsh realization that most of them weren't even born in 1969), the most lovable, unlikely, underdog baseball team to ever win a World Series, and soon we'd adopted their slogan, "Ya Gotta Believe," which went out on the web and began to show up on signs and walls at campaign rallies in every city the governor visited.

It was as if the world were shifting right before our eyes, the ground rumbling beneath our feet. As techies, we had been hearing for a decade that the Internet would radically transform American life. Well, here we were, at the place where all of the things that were supposed to happen in the future were happening right now. To us.

In those early days, you could gather around the computer screen and *see* the campaign beginning to come to life on web sites and in blogs, in e-mails that ricocheted around the country, each hit and blog and message representing a real person learning about Howard Dean and stepping up to do his or her part. You could see it.

Each day in the beginning, Zephyr and I would check first with Meetup.com, which posted the number of people getting together to meet in each area of interest. As we watched our numbers climb steadily—3,000 to 5,000 to 8,000—we quickly left the other presidential campaigns behind us, but there remained a handful of groups whose Meetup numbers were higher than ours, including a surprising number of people interested in vampires, goth, and witches. Especially witches. By early March, we'd passed the vampires and goth fans and soon, all that remained were the witches.

But the witches (actually very nice people interested in Wicca and all sorts of interesting things) were tough. Eventually, we'd get to 11,000, then 13,000, and 15,000, and still the witches remained ahead of us, seeming to grow by whatever it took to keep us in the number-two position.

"Christ, how many frickin' witches can there possibly be out there?" I asked. It wasn't fair. We were only drawing Dean interest from the United States, but there had to be witches who wanted to meet up in every country in the world. Maybe they were recruiting new witches. Or maybe they could magically create *new witches* when we got too close.

"I've been thinking. Maybe it's not such a good idea to pass the witches," Zephyr offered one day. I could see her point.

Finally, in early April we even blew by the witches—and didn't seem to suffer any adverse effects. Now there were 27,000 Deaniacs on Meetup.com—and counting.

The old guard of the campaign watched all this with a kind of careful enthusiasm. The standard measurements of a campaign—fund-raising and poll numbers—were showing some improvement, but there was a lag time and the real effect of this shadow Internet campaign had yet to show up in the old measures. Some people had trouble connecting the freshman-dorm energy of the web room with the business of actually getting traction in the race. In a few months, they would understand what was going on, but initially, they just didn't get it.

Internally, every political campaign is a balancing act between forces trying to steer the candidate and his message. The best campaigns keep these forces—insiders and outsiders; campaign pros and the candidate's longtime staff; message people, money people, and strategy people—in a kind of equilibrium, so that everyone is at least pulling in the same direction. I had worked on campaigns that literally came apart from this tug-of-war, especially when things began going well. When every decision might also determine who becomes chief of staff to the president, the weight of campaign decision-making changes.

Any dissension in the Dean campaign didn't come from those kinds of power grabs, from naked ambition, or desire for pure control. I can say this: Everyone on the Dean for America campaign honestly believed that what he or she was doing was the very best thing for the campaign. Every fight was a sincere attempt to do better and every person seemed to be acting in what he or she saw as the best interest of the campaign. But this can make for more bitter fights than if the division is caused by a thirst for power. In my experience, there is nothing worse in a political campaign than two factions who honestly believe that their plan is in the best interests of the candidate and his campaign—and that the other view will take the campaign straight into an iceberg. Who walks away from that fight? No one. Even when both sides clearly agree there is an iceberg ahead—one side honestly believes that turning to port will clear the iceberg and that turning to starboard is suicide—the other group honestly believes that turning to starboard is the only way out and turning to port is suicide. When there isn't a strong chairman in place to sort things out, the result is usually each side grabbing the wheel long enough to do a zig-zag straight for the iceberg. So there were plenty of disagreements, power plays, and behind-the-scenes drama. No more than on other campaigns, I would say. But on the Dean ship, we added another dynamic that needed balancing: the relationship between the old experienced campaigners who knew the tried and tested things that worked, polling, focus groups, TV ads; and the new kids—idealistic, young, wired people who sometimes wanted to do something new for the simple reason that it was new.

I would spend the next ten months explaining these groups to each other and trying to get each side to see the road that I believed existed down the middle, a road that I was beginning to believe could lead all the way to Washington.

By spring, the burgeoning Web Team would have a half-dozen people—as many as had worked on the entire campaign just weeks earlier—split into two groups, the web side and the blog side. This was the place to be. They worked and chattered and ate and slept at their desks in what we called the bullpen, a long office split by a half wall right outside my door.

The location was no accident.

The geography of a campaign headquarters is important both functionally and symbolically. You can see the internal politics of a campaign—who's making the decisions, who's on the outs—by the way the offices and desks are laid out. Early on, I could tell the people in the traditional campaign offices—like the political and field offices—were somewhat confused (and maybe a bit threatened) by the importance I put on the Web Team—these strange people, some of them just kids, hunched over laptops, headphones over their ears, tapping at computer keys, so that the room always sounded like there was a light rainstorm inside.

To my mind, in some ways, this room *was* the Dean for America campaign, the engine of so much that we were doing. This was the most difficult thing for people to understand, inside and outside. The campaign wasn't in our headquarters. It wasn't in Iowa, or New Hampshire, or Michigan. It was . . .

Out There.

You couldn't see it. And so this was the thing that the traditional media missed, at least early on, the fact that the Howard Dean movement, whatever it was, had taken on a life of its own, becoming a living, breathing organism. Other campaigns in the past had talked about being decentralized, moving decision making out to people in the field, but this was different. We didn't turn the campaign over to organizers in Iowa or in Michigan. We just turned the thing loose. Threw it out there to see who would catch it.

It was Gary Hart's old strategy—but the Internet was like concentric circles on steroids. One day I sat down with a former Hart operative and we were talking about Gary's pebbles in the water, and my old friend said, "Yeah Joe, but with the Internet you guys are raining down pebbles all over the place! And the ripples could be amazing!"

We tried to tell the pundits and reporters early on that there was this groundswell, a wave just beginning to rise, but it wasn't something we could show them, except perhaps, in relation to the number of vampires in the world.

No, at least in February and early March, there wasn't a single rally that we could point to, or an endorsement, or a poll that suddenly showed Dean moving up. There was this sense that people were gathering, deciding to get together. There was just this hum of excitement and activity, the sound of people beginning to ask what was possible.

But it wasn't only the media that was missing and underestimating what was going on out there.

Inside the campaign, we knew something was happening, but you would have been hard pressed to find anyone who would've guessed the extent to which these things we had theorized about were beginning to come true. It wasn't really until the first week of March that we saw the real-world results of all this online talk. It wasn't until the first week of March that we realized we had something, and the rest of the campaign staff saw what I'd been ranting about all those weeks. It wasn't until the first week of March that we knew we had a shot at this, and that we just had one more person to convince of that fact before we made our move:

Howard Dean.

6

NEW YORK

Regime Change, September 11, and Pennies from Heaven

THE MEETUP GUYS were dying.

The idea behind Scott Heiferman's and Matt Meeker's web site was beautiful in its simplicity: Gather people interested in some topic—say Irish Setters—schedule a meeting time—say, the second Thursday of every month—and then find venues in the cities with enough people for a meeting. If there were twenty people, maybe a Starbucks would work; fifty might bump it up to a TGI Friday's.

They had developed a system based on human nature and prior experience to tell them that if forty vampires signed up for the February meetup, maybe thirty-two would really show, and so they had gotten very good at matching up the group to the perfect place.

And then along came the Dean campaign.

We hadn't really paid attention to the January and February Meetups, but as the March events approached, we could feel the stirrings of this . . . *thing* all over the campaign and we were doing everything we could to feed it. With the Meetup link on our web site, all of a sudden in the days before the event the numbers were increasing by the hour.

Based on the February numbers and the people signing up, the Meetup guys had booked Starbucks coffee houses in Los Angeles, San Francisco, and New York for the Dean for America meeting. It was rapidly becoming clear that Starbucks wasn't big enough to hold the fifty or sixty people who were saying they were interested in getting together to talk about the candidacy of Howard Dean.

So a week before the March 5 Meetup, the people in Meetup.com's New York offices were scrambling to find bigger quarters. As soon as they found someplace big enough to hold fifty or sixty people, suddenly the number would jump to one hundred. And they'd be scrambling again. In New York, by Monday morning, two days before the event, *three hundred* people had signed up. By now they had gone through four venues in New York, from a twenty-person place to a fifty-person place, to a hundred-person place, to a two-hundred-fifty-person place. And now they were wondering if they needed a five-hundred-person place. Something bizarre was happening. This wasn't the way their site worked. This was not seven people looking for a place to meet to talk about Irish Setters.

As I monitored all of this from Dean headquarters, I had one of those ideas that seems so obvious, you can't imagine that it didn't occur to you sooner. Sarah Buxton was Dean's scheduler, a bright, twenty-three-year-old redhead, wise beyond her years and nearly as protective of the governor as Kate O'Connor.

On Monday, two days before the March Meetup, I sidled up to her and casually asked where the governor would be on Wednesday.

She checked the schedule. New York.

"Wouldn't it be great," I said, "if the governor stopped by the New York Meetup—"

"No, no, and no," Sarah said. "Absolutely not." The governor's schedule had gotten "out of control" earlier, with too many demands on his time, and she needed to make sure they didn't burn the candidate out with a bunch of small, pointless events. She wasn't about to send him through New York traffic for some bizarre Internet meeting that we had no control over. Her reluctance was understandable; it was her job to make the most efficient use of the candidate's time. And she certainly wasn't the only one who thought that the Meetups would turn out to be online hype. It was easy to imagine no one showing up, or worse: a half-dozen Internet nuts demanding to know Governor Dean's policy on UFO abductions.

I explained to her that three hundred people had signed up for this thing. Three hundred people. No presidential candidate a year before Iowa was playing to crowds of three hundred people without giving away free beer and pizza. Even if it was three hundred UFO nuts, there were still three hundred of them. All day, I begged. I cajoled. I may have thrown a fit. (Scratch that . . . I definitely threw a fit.) She reluctantly agreed, although it was clear that if something had to be dropped from the schedule that day, this would be it.

The day before the event, we sent an e-mail to some of our volunteers who had signed up for the Meetup, saying that the governor was going to try to stop by the meeting. Within hours, the thing had exploded, blowing out the three-hundred-person venue. E-mails rocketed back and forth: *What? He's coming? We hold a meeting and the candidate actually shows up?* In a few hours, the number of people signed up in New York to see Howard Dean was at five hundred. At the web site offices, they had to scramble again to find a place that would accommodate twenty times the number of people they'd originally expected.

This would turn out to be, to my thinking, the first great moment of the campaign—March 5, 2003. I was traveling with the governor that day and I decided to skip out of the event he had scheduled just prior to the Meetup forty minutes early, to see just what it was that I would be walking him into.

The Meetup guys had settled on the Essex Lounge on Manhattan's Lower East Side, because it held 550, which would give them a little bit of room in case everyone who had signed up came—something that had never happened. I'd never been to the Essex, or to that part of New York, and so I was disoriented as I rode in the back of a yellow cab through midtown Manhattan. Finally, the cab turned the corner onto Essex Street.

"Holy shit," I said.

The block was solid people. They formed a line—three, four people deep—coming out the door of the Essex, going all the way down the block and around the corner. *For a political candidate? In March? More than a year before the election?* I just sat there, staring, wondering if the lounge was closed, or if there was a chance they were gathered here for some other reason.

Right then, a New York Police Department car pulled up to the curb and I thought, oh great, the governor is going to pull up and the police are going to be busting people for loitering. Maybe a riot will break out. Yeah, that will be good.

I walked up to the cop, trying to figure out what to say: *Listen, this is either an Internet miracle, or something on the Internet gone hopelessly awry. I guess it sort of depends on how you look at it.*

The cop was talking to the people in line and he turned when he saw me.

"Hello, officer," I said, "I'm with Governor Howard Dean and I can explain—"

"Oh yeah," he said. "I've been talking to these guys about him and I don't know who he is, but do you think I could meet him?"

Fifteen minutes later the governor pulled up and stepped out of the car. His jaw dropped. He had a look of sheer wonder on his face. Before that, I'd always had to remind him to mention after his speeches, that if people were interested in getting involved, they could go to Meetup.com. Most of the time, he forgot. But after that day, Howard Dean would get religion about it, and he'd never fail to say, at the end of his speech, *If you want to do something, go to Meetup.com.* And after that, Sarah Buxton would hold the first Wednesday of every month on his schedule for a stop by the Meetup in whatever city we were in that day—and she would guard it with her life. This was one of our first steps toward an open-source campaign; we decided to let the people choose the campaign events. If Howard Dean was in your town on the day of your Meetup—he was coming. Sarah would make sure of it, and she and I ribbed each other about it for the rest of the campaign.

That day, there were five hundred people inside the high-ceilinged Essex Lounge (more would've crowded in, but the Fire Marshal, conscious of a few recent, horrible nightclub fires, closed the room). Another three or four hundred people were waiting on the sidewalk outside. There were at least *eight hundred people!* Eight hundred people gathering, on their own, on a weeknight on the Lower East Side of Manhattan to have a meeting about a long-shot candidate for president.

We took Dean to the end of the line, halfway around the block, to the last person lined up, and he worked his way forward, shaking every hand—people reaching for him, patting him on the back, encouraging him to keep fighting, thanking him for finally saying the things that they believed.

Sometimes, your own thoughts circle around and catch you off guard. As I stood there guiding a buoyant Howard Dean through that raucous, adoring line, all I could think of was Bobby Kennedy, walking down the street in Harlem or Los Angeles, the crowds forming around him, the hands reaching out, a generation rising up to lift him onto their shoulders.

Inside the long, narrow restaurant, five hundred people were worked into a frenzy. Howard stood on the balcony, looked down at the sea of people and gave them a rip-roaring speech. The Bush administration was steaming relentlessly toward war with Iraq, and the governor decried the president's single-minded and misguided obsession with toppling Saddam Hussein's regime.

"It is time for regime change!" he said. "We need regime change in Washington!" The room went nuts. It was a line that would get John Kerry in trouble with the media weeks later, but when Dean coined it, there were no reporters around, just a roomful of people who had been waiting for someone to have the courage to say it.

I had tried to convince several political reporters to cover this Meetup, but they all passed—and would continue to pass for months. It was one of my biggest frustrations in those early days: this was the most amazing groundswell of support I'd ever seen in presidential politics, and because this thing was happening *out there,* on the Internet, in places where the mainstream media didn't know to look, no one was covering it. The reporters all believed they knew how the system worked, which was the way it had always worked. If support didn't show itself in the polls or—even more importantly in the eyes of the reporters and pundits—in the amount of money a candidate raised, then it wasn't happening.

But, of course, it *was* happening. It was happening in real time, weeks before it would register in polls or in fund-raising. It was happening in living rooms and dorm rooms and classrooms all over the country. It was happening in Los Angeles, where the Meetup blew through its venue and had more than 200 people. It was happening in San Francisco and Seattle, where the venues couldn't hold all the people who showed up for the March Meetup—and there was no candidate at these Meetups—just people who supported Howard Dean and wanted to get involved and do something about it. And it was happening in New York, right before our eyes, where Governor Dean looked at all those people and then glanced over me, and I could see:

He got it.

MAKING A DIFFERENCE

So why Howard Dean?

In some ways, that would become the fifty million dollar question. When the media finally showed up at Dean headquarters, three or four months after this groundswell began, that was the question many reporters wanted answered. Why had this thing crystallized around an unknown, patrician, Democratic presidential candidate from what was, in their view, an insignificant state? Was it his prescient opposition to the war in Iraq? Was it his refusal to speak in campaign clichés and prepared speeches? Was it because we were the first-movers, the first campaign to effectively run our campaign on the Internet? What they were really asking was even more significant: could anyone step in and—using the Internet and other new, emerging forms of communication—replicate our early success in organizing people? Would these same people on the Internet have rallied around anything that took the step of reaching out to them on their level, through their medium?

The answer to all of these questions is yes—and no.

From that first day in Burlington, I had gone out with one message: *Internet, Internet, Internet*. Howard Dean was going to be the Internet candidate, returning power to the American people. When you looked at him, you were going to think Internet and personal empowerment in the same way you thought Vietnam hero when you looked at John Kerry, or Southern optimism when you looked at John Edwards. By summer, when they began to see how much money we were raising, the other 2004 presidential campaigns would belatedly follow our cue and start courting voters on the Net. Most of them missed the boat, although there were some startling successes. In the first quarter of 2004, after Dean dropped out, Kerry inherited much of our online momentum and raised more than half of his $50 million war chest from online donations, including $20 million in two ten-day *Web-a-thons*. But other than raising money, the campaigns generally continued to treat the Internet like a TV with keys (*a Web-a-thon?*) and ignored the ways that this technology actually invites people to be involved, rather than just throwing pictures and slogans at them in the hope that they donate money. Kerry, for instance, still had only 80,000 people on Meetup.com two months after Dean's campaign ended. Dean for America still had 165,000 and *he wasn't even running any more*.

Generally, all of the campaigns made the same sorts of mistakes. Early on in 2003, most of the candidates simply put up static web sites with no place for people to get involved (the exception was Gary Hart, who briefly considered another run for the White House and had, to my horror, a better, more interactive blog than ours—for a while). And even when these other campaigns did venture further onto the Internet, they tended to view it the way most American companies view it, as just another top-down, one-way medium, a place to *get their message across*. They seemed to think the computer was just another box from which to tell the American people what was good for them, what soap they needed to buy, what car they needed to drive, what presidential candidate they needed to vote for.

The other campaigns condescended to the people on the Internet. Like so many American corporations do, in their tone and content, the other campaigns talked down to these people. They didn't engage them or listen to them or invite their opinions. And they made it abundantly clear that they didn't respect the power that these people had.

A story by Ryan Lizza that spring in *The New Republic* referred to the "inordinate amount of time" that I spent using the Internet to create a grassroots campaign. And it betrayed how the other campaigns viewed the waves of people becoming politically involved on the Internet:

> Aides to some of the other 2004 Democratic candidates regard Trippi as a bit of an eccentric who wastes precious campaign time e-mailing obscure bloggers and hanging out with political oddballs at the monthly Dean Meetups. "Some of these Meetup events look like the bar scene from *Star Wars*," says an adviser to one Dean rival.

This became one of our rallying cries. As our web success fueled our fund-raising and then our poll numbers, we wasted no opportunity in throwing the "bar scene from Star Wars" back in the faces of other campaigns, reminding them that they had been too busy to waste their time with the "obscure bloggers" and "Meetup events" that were just beginning to propel us into the race.

So, yes, the other campaigns definitely missed the boat and even pushed away some of their online support. As I watched Bush and Kerry on television marching toward the November 2004 election as if it were still 1976, I was amazed that they still didn't seem to get it.

But in some important ways, it was never theirs to get.

The Internet is tailor-made for a populist, insurgent movement. Its roots in the open-source ARPAnet, its hacker culture, and its decentralized, scattered architecture make it difficult for big, establishment candidates, companies, and media to gain control of it. And the establishment loathes what it can't control. This independence is by design, and the Internet community values above almost anything the distance it has from the slow, homogenous stream of American commerce and culture.

Progressive candidates and companies with forward-looking vision have an advantage on the Internet, too. Television is, by its nature, a nostalgic medium. People will watch twenty-year-old sitcoms and *Behind the Music* documentaries on television as a way to tap into their pasts. Look at Ronald Reagan's campaign ads in the 1980s—they were masterpieces of nostalgia promising a return to America's past glory and prosperity. The Internet, on the other hand, is a forward-thinking and forward-moving medium, embracing change and pushing the envelope of technology and communication.

I do think this gives the Democratic Party a leg up on the Internet ladder—as long as it continues to be the traditional progressive party, intent on moving forward, trying to make the world better. The Republican Party is the command and control party, the very definition of top-down management. That's why what the Bush/Cheney web site calls a blog is really nothing more than a bunch of press releases with no room for reader comment. Nothing seeps up from the supporters; all of the communication is delivered top-down.

It's not that conservative and reactionary forces are unwelcome on the Net. *Everyone is welcome on the Net*. A 2004 Harvard Kennedy School of Government case study of blogging, for instance, reported the perception among many observers that, at least initially, "the strongest voice to emerge from the blogosphere came from the right end of the political spectrum." In part because of what has been called the digital divide (the fact that lower income people are slower to get onto the Internet) the percentage of people online who were registered Republicans (36 percent) in the 2000 election was higher than the percentage of registered Democrats (28 percent).[1]

---

[1] Steve Davis, Larry Elin, and Grant Reeher, *Click On Democracy: The Internet's Power to Change Political Apathy into Civic Action* (Boulder, CO: Westview Press, 2002).

But wherever they fall on the spectrum, Internet activists generally share a few common traits: they tend to be younger, they tend to be unafraid of change—in fact they demand it—and they tend to distrust the mass media to express their particular viewpoint. And so whether it's the Drudge Report or Daily Kos, the most successful web sites and blogs on the Net have about them the unruly whiff of rebellion.

This is how a former wrestler and third-party candidate like Jesse Ventura could use his outsider status to focus the Internet community into raising two-thirds of his campaign donations and driving his victory in the Minnesota gubernatorial race of 1998.

And it's how a Republican Senator like John McCain could use the Internet to raise $6.4 million after his shocking win in the New Hampshire primary. Despite spending a life in government, McCain was able to run as an outsider by aligning himself with campaign finance reform, a populist issue that remains a hot topic on web sites and blogs.

But these campaigns would prove to be limited or primitive uses of the Internet, and in the spring of 2003, most experts still believed this technology was years, if not decades, away from making a real impact on politics in America. What they underestimated was the Internet's ability to grow rapidly, virally, to create *a movement.* What they never understood was that we were not *using the Internet.* It was using us. Although I may have seen this potential before most people, I didn't create this movement. The campaign didn't create this movement. Howard Dean didn't create this movement. In many ways, the movement created the Dean for America campaign.

Certainly, Dean's early opposition to the war mobilized many of these people. But, as Ryan Lizza noted in the same story in May in *The New Republic,* this wasn't a one-issue campaign, like the antiwar campaign of Eugene McCarthy:

> While many predicted that Dean would fade away once the war was no longer a salient issue, there is little evidence that the former Vermont governor's supporters—originally drawn to Dean when he was forcefully speaking out against war in Iraq—are deserting him. In fact the Internet might account for Dean's staying power.

Later, Lizza described a recent wave of e-mails by Dean supporters in which

very few mentioned the war. To the extent any of them dealt with issues, they defended Dean as a New Democrat-style centrist. But most
were nonideological, simply praising Dean for his passion and ability to
bring independents and nonvoters into politics.

As much as they were responding to Howard Dean, these were Americans responding to themselves, to their own involvement in his campaign.
In many ways, engagement became the point of their engagement. The
movement became the focus of the movement.

I think there are many reasons for this renewed enthusiasm in participatory politics. Certainly some of it is a backlash against fifty years of
broadcast politics, which treated the people as if they were nothing more
than fund-raising targets, as points on a poll. Some of it was Howard
Dean, who refused to be packaged like other candidates and was out there
for everyone to see, warts and all, one of their own, a real person running
for president.

But I think the genesis of this participatory movement can be traced
to the terrorist attacks of September 11, 2001. In the wake of the attacks,
most studies showed that Americans were fearful and anxious, but there
was another trend running beneath the fear that people missed. The attacks may have paralyzed many Americans, but others were awakened.
Countless people saw the terrorist attacks as a reason to get involved in
their communities again, in government and in public service—especially
the young.

A November 2003 Young Citizen Survey by the Center for Democracy and Citizenship found that two-thirds of young adults (67 percent)
said they were more likely to participate in politics and voting because of
the attacks. Fifty-six percent said they would be more likely to work in
community service.

These were the people we saw as the swing voters in 2004, not soccer
moms or NASCAR dads, but their kids, all grown up and eager for something to believe in.

And yet, until something like the Dean for America campaign actively went online and asked for their help, these young people *who
wanted to do something* were not being mobilized and didn't feel empowered. The same study found that 62 percent of the young people polled
thought that they could make "little or no difference" in the direction of

the country, 10 percent higher than the number who had felt powerless two years earlier. These young people were also less trusting of the government and its ability to solve their problems. And so, as the CDC study reported:

> Despite their stated intention to participate more vigorously in politics and community life, young adults' civic and political involvement has not increased in recent months. Voter registration and volunteerism rates are lower in this survey than in previous national surveys.

I think that deep down we all really want to make a difference. Deep down, we all want to believe in something and commit ourselves to making the world better. If we did anything right in the Dean for America campaign, it was simply finding new ways to involve people in the process. These young people who wanted to do something but felt shut out of the process found others like them on the Net, and found a place where they could come together—DeanforAmerica.com.

I think in hindsight, we will see this campaign as a crossroads for young people in America, for what might be the last generation raised under the dominant thumb of television, the generation that moves away from the tube and onto the web, from a passive world to an active one.

It was something I began telling people in that spring of 2003, that the most lasting accomplishment of the Dean candidacy wouldn't be felt in this election. Or even the next. But twenty years from now, I am convinced there will be twenty-five or thirty members of Congress who cut their political teeth on the Dean campaign, people like Gray Brooks, who will look back on the spring and summer of 2003 as the moment they realized they had the power to make a difference.

SOMEONE OUT THERE

To this day, we have no clue where the idea came from.

All we know is that it was brilliant.

The day after the March 5 Meetup, an e-mail written by a Howard Dean supporter somewhere went out to all the others connected to Meetup.com. "Listen," the e-mailer wrote, "as of last night there were 22,000 of us Howard Dean supporters at Meetup.com. If every one of us sent the governor ten dollars it would really help the campaign out. But what I think we should do is put a penny on the back end of our donation, so the campaign knows when they get ten dollars and a penny, it's coming from a Meetup person. So give whatever you can give—and a penny."

We were all working in the Dean campaign headquarters, still marveling at the amazing New York Meetup and preparing for the end of the first fund-raising quarter (March 30), when Bobby Clark (the tech guru I counted on to figure out *everything*) came in, looking confused. On our web site, people could use their credit cards to donate money to the campaign and Bobby had been monitoring a surprising spike in on-line donations.

"A weird thing is happening," he said.

We all looked up. "What do you mean?"

"We're getting all these contributions, they're all hitting the server with a penny attached. Twenty-five dollars and a penny. A hundred dollars and a penny. They all have this penny attached."

We had no idea what was happening. This hadn't come from us. This had come from *out there*. Out there, the original e-mail was being picked up and posted on blogs and web sites and the money was flying in. By the end of the quarter, about $400,000 had come in with pennies attached.

We were being taught the most important lesson of the campaign—and the one that most American institutions have failed to learn. They operate under the Field of Dreams paradigm: *If you build it they will come.* And so from 1993 to 2003, they built their web sites and waited for the people to arrive, assuming they'd just appear one day like ghosts out of the corn.

We couldn't wait for ghosts. So we operated under a different paradigm: *If you ask, they'll help build it.* From the moment those pennies started rolling in, it seemed like everything we built, someone *out there* improved on it.

This was also one of the first signs that our efforts to engage people on the Net were paying off in our fund-raising. It was quickly followed by other signs.

One day, I was sitting at my desk, waiting to do an interview with National Public Radio about our Internet campaign, and I happened to click my computer over to the server to see what was happening with our donations. Nothing much was going on, the number was just staring me back in the face. So when the interview started, I looked away from the screen. We talked about empowering people, getting involved in democracy, and in the middle of the interview, I happened to look back at my computer screen.

I couldn't believe what I was seeing. The numbers were rolling like a frickin' gas pump on Labor Day. Click, click, click. Forty-eight contributions. I started surfing the Internet, trying to figure out what was happening. Was there a new poll? Then, after about ten minutes of searching the Net, it hit me.

I looked from the screen to the telephone to the radio and back again. The connection was right there. People were hearing me on the radio, going to their computers and donating to the campaign. The Internet was making it possible for people to register their feedback immediately. After that, we would chart the effect of newspaper, television, and radio stories and be able to predict accurately how much money would come in online after Dean appeared on *Hardball,* or after a story in *USA Today,* and we'd know which media to go to in the big fund-raising pushes.

The campaign fund-raising year is divided into four quarters, and at the end of each quarter the staff is expected to file a fund-raising report with the Federal Elections Commission. This fund-raising number gives reporters and campaign watchers an early picture of the horserace: which campaigns will be contenders and which look like also-rans. At the end of the first quarter, the last day of March, Dean for America posted $2.7 million in quarterly fund-raising, most of it from more traditional campaign sources.

Steve Grossman and Stephanie Schriock had saved the Howard Dean campaign—when it should have been impossible to save. Sure our little Internet thing had raised $400,000 in the last week of the quarter and roughly $600,000 in total for the quarter—but that would have been enough to pay for shutting the campaign down and sending the staff home. Schriock and her finance team, along with Grossman and some others, somehow had raised $2 million the old fashioned way—working the phones for every dollar. Getting people like Don Beyer the former lieutenant governor of Virginia, or Roy Furman, the respected investment banker and former DNC finance chairman in New York, or Rick Jacobs and Steven and Mary Swig, the anchors of our campaign in California to sign-up and help raise the money that kept us alive. We needed time to give the Internet and our growing blogging community a fighting chance. They raised the lifeline money to make it happen—and got none of the credit.

I was both ecstatic at what they had accomplished, and pissed that they had raised just enough to make sure that I spent another winter in freezing Vermont. I let Schriock know what I thought of that prospect by giving her the nickname LBSOS—Low-Balling Sack of Shit—for swearing to me throughout the first quarter that we would be lucky to raise about half of what she and her crew actually brought in. At one point, Schriock had so convinced me of our dire financial position that I made sure Kathy and I maxed out with $2,000 contributions each. Then to make my pain and humiliation even more profound, on my birthday, Schriock and Kathy conspired to convince me it would be a great idea for me to sit in a dunking booth and let the campaign staff (the people I yelled at every day) throw at a target that would flush a toilet full of ice cold water on me. Then they put a picture of the whole damn thing up on the BlogforAmerica and asked people to contribute to the Dean for America campaign as some kind of sick twisted birthday present—to me! There was nothing like sitting in a dunking booth for hours in the freezing drizzle while a bunch of adrenaline-pumped staffers got their revenge. The only pluses were that we raised six hundred bucks and that Governor Dean was on the road that day—he has a wicked fastball and who knows how he would have reacted to the chance to put me where the fishes sleep.

I had already gotten even with Kathy for this trick by marrying her. To Schriock I could only add another L to her nickname—*Lying* Low-Balling Sack of Shit—for her mind-numbing insistence to me that there was no way we were going to put over $3 million together in the second quarter—when she knew full well we were going to do at least *$4 million*.

Even with Stephanie's sleight-of-hand, we were well behind the leaders, the two Johns: Edwards, who had raised $7.4 million, mostly from fellow lawyers, and the obvious front-runner, and Kerry, who had topped $10 million.

To the media, we were in the same position we'd always been in, underfunded and outgunned—too little, too late, and too liberal. Reporters wanted to know if the governor would drop out of the race once the primaries started, or if he planned to run a symbolic campaign, getting his ass kicked until the very end.

In Burlington, we were just beginning to reap the benefits of our online strategy, but it would be several weeks before it translated in those traditional measures of polls and cash. So we were brainstorming about what other innovations might help us reach these voters that we'd just begun to find on the Internet—and to bypass the traditional media, which had decided months earlier that the long shot Dean for America campaign was unworthy of coverage.

I'd had an idea for some time, a notion that could essentially turn the campaign into a reality television program—send a camera out with the candidate every day to film the rallies and debates, everything going on behind the scenes and on stage. No secrets, no backroom dealings—open up the campaign and let the people see inside it, a running journal of a campaign, an all-access video blog.

This is the opposite way that political campaigns generally function, of course. Most campaigns do everything in their power to control every element of the candidate's image and message, from the clothes he wears to each word out of his mouth. But we had a candidate who didn't like to have his image or his words packaged—he delivered only two written speeches during the entire campaign—and whose very identity was tied up in the fact that he was his own man.

That spring, we launched HowardDean.tv, using cool technology from Wavexpress, a company I had once consulted with (and lost as a client the day I took the Dean campaign job). HowardDean.tv was a web site that ran video of speeches, campaign events, and, best of all, video clips shot by Dean supporters around the country. It was to be a fully dedicated TV channel, but because we were still broke that spring when we hatched the idea (we couldn't even afford airfare to send a cameraman out with him) it never quite lived up to my expectations, though by the end of the campaign, 50,000 people were catching some of the 24 hours of Howard Dean video clips and speeches that aired each day.

The best things on Dean.tv turned out to be the pieces made by our supporters—commercials and testimonials and over-the-shoulder glimpses of rallies and other events, and among these were occasional glimpses of the kinds of powerful, unscripted "reality" moments that I had hoped to get on Dean.tv.

These moments were incredible and only deepened my belief that someday a campaign will do this to perfection, will understand that people—especially the young—see through the slick packaging of TV and prefer seeing the world raw and unscripted, not as it's imagined in a Madison Avenue conference room or a Hollywood studio.

My favorite moment on Dean.tv came when the governor arrived to speak to about twenty young voters on a college campus. The amateur-run camera caught up to him, arriving late, volunteers running up to brief him and guide him to the event. With the camera jiggling over his shoulder, he entered the door to a campus building and someone stopped to introduce him to the two students who had organized the event.

"It's so great to meet you," one of the students said. "I'm so excited that I'm skipping my final to be here."

The governor stopped in his tracks and turned to face the kid. He looked like he was going to have a heart attack. "What? What? You can't do that." For just a moment, he didn't look like a candidate for president of the United States. He looked for all the world like a worried parent. "Look, I don't want you doing that for me. You need to get back to class right now."

It was an amazing, sincere moment and it made me wish we could have pulled off the full Dean.tv experience. It would have been the epitome of the campaign I wanted to run, driven by the people, taking out the filter of ABC and FOX and CNN and allowing the viewers to not only choose what they want to see, but to produce it as well.

But in the spring of 2003, we just didn't have the time, the money, or the people to do everything we wanted to do. It's funny to read about our inevitable success—as if we just rolled out the car and watched it tear around the track. I'm really proud of everything we did, but I know how much pushing uphill we did. Sometimes I find myself imagining all that we could have done if we hadn't started two years late, with no money and a tenth the number of people we needed. Hell, if we'd had one more year—who knows what we might have accomplished.

HOLD THE LINE

One day, John Kerry just started whacking us.

That might have been the first real sign that we were making some headway in the outside world—the world beyond the borders of the tight Internet community that we'd been building. At the end of April, the Kerry campaign suddenly turned on its heels and came after us. And, in the process, they helped make us.

From January to June I couldn't get the media to cover what was happening in our campaign, but something in Kerry's polling numbers must've tipped him off to the deep rumblings of our grassroots movement, because all of a sudden, the shells started landing in our backyard:

> Howard Dean's stated belief that the United States "won't always have the strongest military" raises serious questions about his capacity to serve as Commander in Chief. No serious candidate for the Presidency has ever before suggested that he would compromise or tolerate an erosion of America's military supremacy.

I stared at the press release from Chris Lehane, Kerry's communications director and an expert on opposition research. This was a two-week old quote, taken out of context from a *Time* magazine story about searching for diplomatic solutions to world crises. What the hell could this kind of low blow mean? We weren't a threat to Kerry. He was still killing us in the polls, in fund-raising, in all the traditional establishment measurements. Every campaigner knows you don't pick a fight with someone you're beating. Especially in April.

But that's exactly what they'd done. When a staffer from the other campaign attacks your candidate, the candidate can never lower himself to respond. So our communications department has to respond to their communications department. The only problem was that, in April, we didn't have a communications department. I was the communications department. So the retaliation had to come from me:

> The statement by Senator John Kerry's campaign is absurd. As Commander in Chief, Howard Dean will never tolerate an erosion of American military power—Trippi said if Kerry supports Bush's approach to foreign policy "then John Kerry is running for the nomination of the wrong party . . ."

A few minutes after my response went out, Lehane fired right back with another press release, saying that my statement was "a non-answer," and—the adrenaline flowing—I fired back that Kerry had given Bush a "blank check" and on and on we went, the tag line on each press release getting longer and longer: "In a response to Joe Trippi's response to Chris Lehane's response to Joe Trippi's response . . ."

I hadn't had a good political fight like this in years and it was alternately invigorating and exhausting. With their 10-to-1 advantage in money and staff, fighting the Kerry campaign was like taking on the neighborhood bully. In between each salvo, I raced down to my car and cranked the stereo on Peter Gabriel's song "San Jacinto," yelling out the chorus over and over: *I hold the line—the line of strength that pulls me through the fear—I hold the line.* Then I'd run back upstairs to see what Lehane had fired back at me. Back and forth we went, until finally, I sent this message, and effectively ending our "feud":

In a response to Chris Lehane's response to Joe Trippi's response to Chris Lehane's response to Joe Trippi's response to Chris Lehane's statement, Joe Trippi says, "Who the hell is Chris Lehane?"

The press laughed and Lehane (thank God) didn't respond. I had nothing left at that point. I was exhausted.

The thing about an insurgency campaign is that you are always chasing. You're always running from behind. You start behind and you finish behind. With a front-runner, you have all the money you need. You start with a staff of a hundred people, the best schedulers, the best speech writers, the best field organizers; you've got more people in the communications department than Howard Dean had on his whole campaign. I know. I worked on the Mondale campaign and it was a cruise ship compared to the Dean campaign.

In the beginning, I was handling the duties that a press secretary would normally have, did field work, fund-raising, political directing, and with Howard still unwilling to ask his wife to campaign (when she finally did stump for him in Iowa, she was amazing), I was even a surrogate speaker at some events—all jobs that the bigger campaigns had staffs to do. My normal campaign manager day started at eight in the morning and was filled with meetings with labor leaders and politicians, decisions about media buys and campaign events and message and strategy and all the things that make up a campaign until my day began winding down, around 10 P.M., when I'd get on the Internet, read the blogs, and post myself. At about midnight, I'd answer e-mails for a couple of hours. With the Web Team I would work on e-mails to our supporters, often not getting the final draft done and approved until two in the morning. Then I'd get a few hours of sleep and get up and start over again. And it wasn't just me. Everyone on the campaign was working that hard. Everyone was doing the work of ten people. There was always someone trying to catch a few minutes sleep on my couch. In the web room, people slept at their desks. Eventually, Nicco just stopped going home, pulling his hat over his eyes and slumping at his desk when he needed to sleep.

It wasn't that we didn't try to get experienced campaigners on-board that spring. I called more than a hundred people—field organizers, press people, message people—anyone I could think of. Most flatly said no and told me that I was crazy to work for such an underdog. Sometimes, when I would get a heavy hitter to think about coming to work for us, Dean and his close staff would balk at spending that much money before we knew if this thing was even going to fly. My one real regret in terms of staffing came when I asked Ken Bode, someone I had known as NBC's national political correspondent in the 1984 and 1988 presidential campaigns to head up our press operation. Ken had always been tough on me during those campaigns—but I remembered his political smarts and always admired that take-no-bullshit side of him. I badgered and begged and I think he was almost crazy enough to come on board—but he had just signed on as the dean of journalism at DePauw University, so he reluctantly (I think) took a pass.

Still, alongside the campaign rookies and eager techies, we were beginning to attract some talented young pols, like the tireless Paul Blank, who liked to think of himself as something of an expert on what women voters wanted (namely, him); and the quick-witted, irrepressible Tricia Enright, who came on as communications director and crackled in the part like Katharine Hepburn in one of those old forties office comedies.

But even though we were attracting good people to the campaign, we seemed to fall further behind each day. The first minute I walked in the door, I thought: *Oh god, we need thirty people and we only have seven.* And the thing about an insurgency is that you won't catch up. Ever. You think you need thirty people, and you get to twenty-eight and then your candidate does something like the DNC speech—whacks it out of the park—and you're thinking, holy shit, now you need sixty people. You get to fifty-eight, and you realize you need eighty. At seventy-eight, you need a hundred and fifty. And this never stops. In an insurgency campaign, you're always outrunning your supply lines. You're always spending money you don't have, relying on people who aren't there, bailing water out of a boat that is sinking under your feet. You don't manage an insurgency; you survive it. You ride it like a surfer on a tsunami. You can never go home at night without leaving more work on your desk than when you arrived. Managing an insurgency campaign is like one of those dreams where you're falling, and you just keep falling, and falling.

Except when it works. Then it feels like you're flying.

7

JUMPING FROM A FIFTEEN-STORY BUILDING

Puerto Rico, the Russert Primary, and Overlooking Snail Mail

"HOW'S IT LOOK?" I asked nervously.

"I'm not there yet," Howard Dean said as he walked from the car toward a campaign event in Seattle.

I had good reason to be nervous. That early in a campaign—mid-May 2003, still eight months before the Iowa Caucuses—it was notoriously hard to gather people for a campaign event. A decent-size crowd might be a hundred people. The biggest event I'd seen for the other campaigns that spring was a breakfast in Iowa with five hundred people and free food supplied by John Kerry. Here we were trying to get a crowd with no food at an event in a state that wouldn't be on the table for months.

"I'm walking up the steps, Joe."

Every time your candidate goes out on the road you hold your breath, half afraid he's going to be speaking to himself in an empty room. The biggest campaigns can usually build a crowd with their support on the ground—hit up the unions or churches—but we had no unions or churches. Hell, we had no official campaign organization on the ground anywhere, except in Vermont, and a little bit in Iowa and New Hampshire. For us, every campaign stop was a small leap of faith.

"I'm opening the door."

But we did have the Internet. Meetups were spreading, and that was good, but it was also only one day a month. A campaign is ten years of events, meetings, and rallies compressed into one year's time. So whenever we had an event, we turned to the best organizing tool we had: the Internet. In Seattle, for instance, we had sent out about two hundred e-mails to our supporters in the local Dean organizations, hoping they could help us raise a decent crowd, maybe even get a few hundred people without having to resort to free bacon.

"Well?" I asked.

There was a pause. "Good Lord, you're not going to believe this," the governor said. "There have to be a thousand people here."

There were, in fact, *twelve hundred* people waiting for him at Town Hall in Seattle that day. They had to close the doors and turn people away. And an event the next month in Austin, Texas, was even bigger. We sent out four hundred eighty-one e-mails to Austin supporters, and on June 13, more than three thousand showed up. Three thousand people for a former governor of a distant state, seven months before a single primary or caucus vote was cast!

What had happened—what would happen the rest of the campaign—was that in Seattle, those two hundred people became little campaign managers, putting up signs and posting flyers, arranging media, and passing the word on to their friends. In Austin, those four hundred eighty-one people had decided, on their own, to leaflet every Latino neighborhood in Austin.

We couldn't have done that from Burlington. We would never have known to do that.

If we hadn't realized it before, by May and early June of 2003, it was becoming clear: This was no ordinary campaign.

This is what I meant when I said later that the biggest myth of the 2004 election was that Joe Trippi was managing Howard Dean's presidential campaign. *They* were managing the campaign. It wasn't headquartered in Burlington; it was *out there.* Anything we could do, *they* could do better. We'd put a new campaign flyer up on the web site for people to download, and in a few minutes, we'd get e-mails with new, improved versions of this same sign. We'd post those on the blog, and someone else would make improvements on the new flyers.

Early on, we had gotten some guidance from MoveOn.org, a pioneer in using the Net to raise money and awareness for political causes. Formed in 1998 to battle the Clinton impeachment and, later, to raise money for ads against the Bush administration and for other liberal causes, MoveOn didn't support the Dean campaign, offering its help to all nine of the Democratic Party contenders.[1] But we were the only ones who accepted the offer. And so Zack Exley from MoveOn came over to show us what had worked for them. But the most valuable lesson was one we learned on our own: If you pay attention to the community you're building, then the community will step up and do the work. Every day, we saw a new example of this.

One day, we put up fifty state signs on our web site—"Iowa for Dean" and "New Hampshire for Dean"—from Alaska to Wyoming. The signs were downloaded 87,000 times that day. But three minutes hadn't passed before an e-mail came from someone in Puerto Rico: "Hey you guys screwed up. You forgot Puerto Rico." They were right. Even though Puerto Rico is not a state, Puerto Ricans still get to vote. So we cut and pasted and made a Puerto Rico sign and two minutes later it was up on the site and we immediately got eight thank-you e-mails from Puerto Rico. But we also got a posting on our blog from a guy in London saying that he'd love an "Americans Abroad for Dean" sign, so we made that sign, too, posted it, and immediately got a thank-you note from a woman in Spain. All of this was instantaneous—almost no separation between planning something, disseminating it, getting feedback, and improving on it. We'd post on the official campaign blog that we were thinking of creating what we called GetLocal tools—software to help people find and organize Dean events in their own area—and fifty people would respond with suggestions on how to make our idea even more effective. Some people even wrote their own software and sent it in.

Most of these early interactions with our online supporters were these tiny moments—a flash of inspiration here, another there. But these flashes were going off instantaneously, hundreds of times a day, like cameras at the Super Bowl, and these flashes were not just between campaign headquarters and our supporters, but back and forth between the supporters too, all

---

[1] In an online "election" between the nine candidates, Dean won 44 percent of the 317,000 MoveOn.org members who voted, almost double the support of the next highest candidate.

through the crowd. Our entire online community seemed to be lit up by the sparking of these ideas. And it wasn't the top simply allowing ideas to rise up from the bottom. Ideas moved up and down, side-to-side, and around in circles, in three dimensions. A woman in Pennsylvania sells her bike for $75 and donates the money to Dean for America, then mentions it on a blog and before you know it donations are coming from all over the country with the note: *I sold my bike for democracy.* We'd get hit by another campaign criticizing some innocuous thing that Dean had said, and the Google Monkeys would break out their search engines and come to our rescue with three former presidents who'd said the same thing. Two Washington, D.C., students would decide one day to form Generation Dean as a way to get young people involved in the campaign and within months, it would grow to 23,000 members in 1,100 chapters across the country.

The multiplying effect of this movement didn't only manifest itself building supporters. Even our ideas were growing exponentially. And while we'd been saying all along that this could happen—this groundswell of support building the campaign at hyper speed—I'm not sure any of us believed it *really would.* But after those campaign stops in Seattle and Austin, as the Meetup numbers swelled past 24,000 people, as the campaign grew more innovative and less tethered to Burlington, we could see it happening before our eyes. And we began to talk about what it meant— the campaign's next step.

Then, one day in mid-May, I sat down to try to capture the energy of what was happening in a post on the DeanNation blog:

> Three months ago Howard Dean was a political asterisk; today he has become such a threat to the frontrunner, and evidently, at least a few others, that they have taken every opportunity to twist his words, and distort his centrist record in what has so obviously become a desperate attempt to stop him before he can't be stopped.
>
> They are trying to stop the Perfect Storm.
>
> It is a storm that has never happened before—because it could not have happened before. The forces required to come into sync were not aligned . . .
>
> First the storm requires thousands, perhaps millions of Americans to become actively involved in determining the future course of our country. But how do these Americans find each other? How do they self-organize? How do they collaborate? How do they take action together?

For the first time . . . the Internet makes this possible . . . (A)mazing tools that have emerged. The Blogging community has grown with readership in the millions, but the one question that still needed to be answered was this: Could the Internet be used by millions to take action off-line. . . . In 4½ months Dean Meetups members have grown from 432 to over 24,000 and still growing.

The other thing that is needed is a campaign organization that gets it . . . every political campaign I have ever been in was built on a top-down military structure. . . . This kind of structure will suffocate the storm, not fuel it. . . . (T)he important thing is to provide the tools and some of the direction . . . and get the hell out of the way when a big wave is building on its own.

Which gets me back to the Perfect Storm. People ask how are you going to win the nomination, or beat George Bush when he is going to have all that money?

My answer would be—that never—until now—would there have been any hope of 1 million Americans contributing $100 each to take back their country and promote a common vision for the future of the nation. Maybe it will be 2 million who contribute $50. But the Internet makes that possible. . . . The tools, energy, leadership and the right candidate, are all in place . . . the wind is getting stronger, and the waves are getting higher, the Perfect Storm is building . . .

This was the first time, according to Nicco and others, that they had been able to step back and see where this thing was headed. The Perfect Storm post would turn out to be a kind of statement of intent for the campaign—part operating instructions, part call to arms. We'd been talking about these things quietly around the office, but now we began saying them out loud, to each other.

And then, one day, I sidled up to Howard and told him what we had all been thinking.

"The people are coming to this thing," I said. "And whatever we do, they take it and make it better. It's their campaign now. We're at the point where, if this is going to work, it's going to be because of them. All we have to do now is have faith in them."

The governor nodded in full agreement.

"We've all been talking about it, and it's like this," I continued. "It's like we're standing on top of this fifteen-story building. All these people have gathered. Now . . ."

I stepped in closer. "What we have to do is jump. And trust them to catch us."

He stared at me for a while and I could see the idea working in his mind. "You're absolutely right," he said. "I can see it. But do we have to be that crazy about it?"

This was for me, a classic example of the cognitive dissonance Howard Dean had about both his candidacy and his campaign.

Here was a brilliant governor with a rare combination of vision and dead-level honesty, one of the few people in the country with the ability to lead a full-scale populist movement to take back the government. And yet also, here was a guy who had learned that you succeed in government by being measured and logical. His instincts were those of a political centrist, not a flame-throwing rebel. Throughout the campaign, I could see him wrestling with the consequences of the kind of insurgency that he suddenly found himself leading. After his triumphant speech in front of the DNC, for instance, when he called out the Democratic Party for appeasing the Bush administration, one of the staffers mentioned that Senate Minority Leader Tom Daschle was angry about the speech. I could see the old conciliator in Dean come to the surface. "Should I call him?"

"Governor," I said, "if you do, you might as well put him on your speed dial. Because you're going to be calling him every day of this campaign."

When I finally left Dean for America, I laughed at the news stories that said Howard and I didn't get along, or that our differences kept us from working together closely. I have a huge amount of respect for Howard Dean. He is one of the most extraordinary men I've ever known. There was never any tension between us personally, even on my last day in the campaign. We were able to joke with each other and make fun of each other, and I think we saw the campaign the same way: it was about empowering the people who had lifted Howard onto their shoulders. I've worked for candidates before who didn't "get" their own message, but Howard Dean got it. As our pollster Paul Maslin wrote in a story for *Atlantic Monthly* after the campaign:

> Howard Dean and Joe Trippi, although their work styles were such that they rarely spoke to each other (and they would ultimately part ways), were nevertheless on the exact same tactical page most of the time—if

not always for the same reason. And when they were, bold action usually ensued.

But Howard was also driven by loyalty, distrustful of outsiders, and cautious about spending too much money on a quixotic campaign. And so, I never felt as though I *completely* had his trust, or the trust of his closest aides. I never quite penetrated the inner circle that Howard Dean kept so closely around him. I don't blame Howard. He didn't know me. Once, when someone asked what I would've done differently, I joked that I would have gone to high school with Howard. Because when the campaign began to lift off the ground, I was frustrated that I couldn't always do what I thought needed to be done to keep it aloft.

For instance, when it became time to step up our television ad buying, he was reluctant about committing to spending the money until he had a contract telling him exactly how much my company, Trippi, McMahon, and Squier, would charge to produce the ads.[2] To me, this was quibbling over minor details at a time when we needed to be acting boldly. "I've been here since January," I said to him. "And I haven't been paid a dime. Governor, you should know by now that I don't care about the money. This isn't about money for me. This is bigger than money."

Later that day, Dean called my partner, Steve McMahon. "I just learned something important about Joe," he said. "He's doesn't care about money. I can't have someone who doesn't care about money handling my campaign's finances." And that's why I never had control of campaign spending. I was supposed to come up with ideas, but if I wanted any of them implemented, I'd have to go through that inner circle.

This was the only real conflict between Howard Dean and me, and I think it was also a reflection of the conflict that burned within him: the battle between the idealist and the pragmatist. He was someone who sweated the details; I was someone who thought focusing too much on details could sometimes make you miss the big picture. This was an honest difference in our temperaments, but most of the time we drew great

---

[2] Later, the *Washington Post* and others would question the $7 million spent by the Dean campaign on television ads, which were produced by Trippi, McMahon, and Squier. Most of that money went to Iowa television stations to buy ad time. We received a 7 percent commission for making and selling those ads—half our normal rate—and so my third of that was about $165,000, my entire fee for thirteen months' work, since I took no salary as campaign manager.

creativity from the contrast in our personalities. And if I represented the governor's wild-eyed idealist side, perched on his shoulder, telling him the only way to win was to burn down the old corrupt system, then Kate O'Connor and Bob Rogan (the deputy campaign manager) were the pragmatic voices, constantly reminding him that if he burned down the castle, he'd have no place to rule. Most of the time, we reconciled those two halves of the campaign (and those two halves of the candidate) but that division would always be there, from the first day I arrived in Vermont, until the day I left the campaign.

"Joe's a genius," Dean generously said to a reporter from *USA Today* at one point. "The only problem is, I'm not used to having staff that I have to rope in. Usually it's the other way around—the staff has to rope me in. I finally met my match."

COME TOGETHER IN COMMON CAUSE

June was the month. We all knew it. June could make the campaign, or it could end it once and for all. June was when we pushed this rickety thing down the runway to see whether it would either fly or fall.

Back in January, when we'd drawn up our Internet strategy, this was the moment we had hoped to get to: June, the end of the vital second fundraising quarter. It had seemed like a horrible long shot, but if the campaign was somehow viable by June and if we'd figured out a way to use technology to grow our grassroots movement exponentially, then there was no telling how far we could go.

But as June finally arrived, there was something missing. The Internet engine was chugging beneath us, but on top, the campaign seemed to be flattening out just a bit. Our message was becoming lost and we were spending too much time talking about the things all the other campaigns talked about—money and polls and attack ads and press and Iowa. We were being a little bit careful, too cautious. Too earthbound. It wasn't anything big, just the slightest shift in tone. But it was dangerous.

And I wasn't the only one who thought so.

As I sat down in late May to write a memo to the governor about my concerns, I got a call from my old friend Pat Caddell, formerly the Democratic Party's *enfant terrible,* now a writer and political commentator in California. Even though he'd been away from hands-on campaigning for years, in my opinion, Pat was still the most brilliant person in the world at taking the political temperature and the mood of the electorate and refining it into a message. The Democratic candidates had just finished their big debate in South Carolina—the Collision in Columbia—and Pat and I agreed that Howard was really beginning to connect with people in a profound way with lines that we'd helped him write for his closing: "The biggest lie people like me tell people like you from stages like this at election time is that if you vote for me, I'm going to solve all your problems. . . . The truth is, the power to change this country rests in your hands, not mine."

The line had started with something the governor said to me before the debate about the truth being that people in this country really had the power to change things. I called Pat and we talked about it and then Pat said, "Yeah, yeah that's it—but Joe, if that's the truth, what's the lie?" So, minutes before the debate, Howard and Steve McMahon and I were sitting in the holding room when the governor just blurted out that the lie was easy—it was politicians telling people, "vote for me and I'll solve all your problems." This became the signature closing of every speech the governor gave from that point on: "You have the power!"

"He's starting to get it," Pat said, "but he's got to go all the way. He's got to go to that point where he shows that he's really different than all those others guys."

"That's right!" I said.

"Look," Pat began, "I'm working on this memo—"

"I'm working on a memo," I said.

So I flew out to L.A., where Pat and I worked together until we had a nine-page memo outlining how to take the campaign that last mile, to the only place it could go. Actually, it was more manifesto than memo, the idealistic rantings of two old campaign warriors who had seen too many politicians aim too low, and who wanted, with all our hearts, to see this one rise above all the bullshit.

"Memorandum. To: Governor Howard Dean," we began. After some debate we left the "From" line blank (there were enough people in the campaign who would be scared of a memo like this coming from me; a memo from me *and* a notorious political bad boy like Pat Caddell would have been too much). In the "Re" line we typed: "Definitional Moment."

The campaign has gotten to a place no one ever thought it could get to.

A confluence of your passion, events of the country, the mood of the voters, and the conjunction of history have produced yet another

moment that is with precedence in American history—the transformation of American politics.

It began with Andrew Jackson who transformed America into a
Democratic Republic, then to Lincoln who saved it, and to the pop-
ulist/progressive movement of Teddy Roosevelt, and Woodrow Wilson
and then to Franklin Roosevelt and the New Deal.

This is another one of those moments—the place where the future
happens.

You have felt this—you know that something bigger is happening
here beyond conventional politics. It is what happens every time you
tell the people that the future of our country rests in their hands—and
not in yours. The room goes silent and you feel the hunger in them and
the frustration within yourself to explain something that you have yet
to find the words to express . . .

This is the thing you must recognize—the thing above all others
you must understand. This campaign is not about you—it may have
started out that way but you have touched something more powerful
than any other force in our nation's history . . .

You have touched a nerve of unvanquished hunger, and almost
limitless need to transform our country . . .

From there, we spelled out where the country had gone wrong, how
we were under the tyranny of *transactional politics*, in which leaders
were expected to do nothing more than "negotiate deals—and grease the
wheels of Washington." We explained how this had led to our nation
being held hostage by special interests. And then we showed the path he
needed to take the rest of the campaign: leave transactional politics behind for *transformational politics;* define himself as the one candidate
who knows the road back from this cynical place, back to a kind of politics that could heal a nation.

We are going to draw on America's history and her traditions.

We are going to cloak ourselves in the mantle of our nation's
greatest leaders at our nation's greatest moments, Republican and Democrat—for this is not a Democratic Party Campaign—this is an
American campaign . . .

The rest of this campaign will be the process of asking the American people to participate once again in their common future—to meet
with you and other Americans across this nation—to come together in

common cause—in town meetings and town halls—to forge a new American Century from the bottom up—from the people of this nation will the greatness of America rise up to rebuild and reclaim all that has made us what we are . . .

The time is now to lead this nation, not run a tactical campaign. And to make sure as Lincoln promised—"that government of the people, by the people and for the people shall not perish from the earth."

You have the power to make the American people realize that they have the power to make it so.

For Pat and me, it was more than a memo. It was a plea for this one candidate to finally rise above the shit, above this system that we'd both devoted too much of our lives to, and that we both knew was rotten to its core.

There must have been something in the air. I had just gotten back from L.A. when my phone rang. It was Joe Costello, the young political genius who had developed Jerry Brown's groundbreaking 800 campaign. "Hey, I think Dean is close to getting it," he began. "I've been writing this memo . . ."

"Get in line," I said. And then I invited him up to Burlington to help the campaign work on its message.

In the second week of June, I gave the governor our team memo.

For days he didn't say a word.

I should not have been too surprised. Howard Dean is not someone taken to emotional rallying cries, not the kind of person who would respond to a fiery call-to-arms. His amazing success as a governor was due to the fact that he is a serious, deliberate man, a physician who doesn't make a decision until he has a full diagnosis. Finally, several days later, he came up to me, his face knit with deep contemplation.

"You do understand," he said, "that in some ways, I'm the most transactional person on the planet." We stood across from each other for a minute, and then he smiled that grin he got when he was ribbing me, and walked off. But we weren't done. Over the coming weeks, I could see him thinking it through, the way he processed things, digesting the ideas, and imagining . . .

And then, suddenly: finals week. The last week of June—the most pivotal point in the campaign so far, when we would have no choice but to finally step off that building and see if anyone was there. If we had a good week, the nontraditional base that we'd built through web sites and Meetups and blogs would show up in the real world and catch us. If not, the Dean campaign would be a stain on the sidewalk.

That was also the week we were announcing Howard Dean's candidacy at an event in Burlington (this is a formality, traditionally done months after the campaigning actually begins), and hoping that the press would finally tune in to what was happening—this national groundswell that they had so far ignored.

As if that weren't enough, to kick it all off, the governor was beginning the week by going on *Meet the Press,* where he'd be interviewed by Tim Russert, the thoughtful, impeccably prepared, and exceedingly influential host. Among the political class, there is no more respected (or feared) pundit than Russert. Pols talk among themselves about winning "the Russert Primary," the early round of Russert interviews that can build major league momentum for a candidate or can reveal him to be a minor leaguer on a thirty-day contract. The conventional wisdom around Russert was that the first interview or two with him was batting practice, when he'd give your guy the soft stuff so he could show his strengths. Dean's first two interviews with Russert had certainly gone this way—friendly, substantive chats about health care and Iraq and the Internet and the future of the Democratic Party. Some people credited Dean's first interview with Russert a year earlier with launching his campaign.

But now, it was a year later and Dean's presidential candidacy was no longer theoretical, no longer just a bunch of talk. This time around, we knew, Russert would bring the heat. He would break out the wicked curves and the knucklers and the spitballs. We knew we had to prepare Dean for that third interview or Russert would bury him.

And so, a few days before the Russert interview, Bob Rogan and I flew to Minnesota, where the governor was campaigning, to begin prepping him for what would be the most important TV moment of his campaign to that point. But when we landed, we found out that the governor was headed back to Vermont. His teenage son had been arrested for breaking into a country club and stealing some beer. I have the deepest respect for the way Dean handled this. He stopped campaigning entirely and flew home to be with his family, to deal with the problem, and make sure that his son knew that this wasn't going to affect the campaign, that it was just an issue between a kid and his dad.

For the next forty-eight hours, the cone of silence descended on the campaign. We patiently waited for Howard to sort out his private life. But now we had a dilemma. Russert was now T-minus one day and counting. Which meant there would be no time to prepare. The choice we had now was that classic campaign moment of picking your poison. We could (1) let an unprepared Dean go on *Meet the Press* and, more likely than not, get pummeled or (2) cancel, using the perfectly good excuse that the governor had to take care of this personal issue, but in the process, drawing even more attention to what would otherwise fade as a minor story—the governor's kid screwing up a little.

We debated it and weighed our two bad choices (a similar exercise might be considering which you would rather hit yourself in the head with—a hammer or a bat). It was fifty-fifty. Whichever path I suggested, I could either come out looking brilliant, or looking for a job.

In the end we decided to go.

Right away, you could see that Russert was in mid-season form. The subtext for the interview seemed to be the question of whether Howard Dean was "presidential material." At times the show played less like a news interview and more like a pop quiz by a professor who thinks his student has been slacking. Russert asked how many active duty personnel were in the U.S. military and how many were stationed at that moment in Iraq. When Dean gave accurate but approximate numbers, Russert chided him for being unprepared.

"For me to have to know right now—how many troops are actively on duty in the United States military—it's silly," Dean said.

"No, no. Not at all," Russert came back. "Not if you want to be commander in chief."

It only got worse from there. It was an unmitigated disaster, a head-on train wreck. The *Washington Post* called it "embarrassing." The *New York Times* called it "a debacle." But I think they were just being nice. Another New York newspaper called it "perhaps the worst performance by a presidential candidate in the history of television."

Well, I thought just minutes into the interview, this is going to be it. We're road kill now. At least I would get to go home to my farm in Maryland and finally get some sleep.

A few minutes later, Nicco stuck his head in the doorway of my office. "Are you seeing this?" While every pundit and cable commentator was on the air reading our obituary, something very different was happening on the Net.

First, the blog was going crazy. Not just ours, but Dean blogs all over the country were humming with even more support for the governor, praising him for "slamming Russert as being inside the Beltway" and criticizing Russert for being an attack dog. Even among those Dean supporters who acknowledged that the governor had looked bad, they said it wasn't a big deal. People could see through Russert's questions. And anyway, they said, the campaign will get over it. Relax. It's just one interview, six months before the first vote.

Our supporters were not running for cover. They were running to help.

But even more stunning was what was happening with the online donations. On normal Sundays, even if we got good press that weekend, we'd be lucky to raise $3,000. That Sunday, *$90,000* poured into the campaign.

It was one of the most humbling and illustrative moments of the campaign for me. Here I had allowed myself to fall into the *old political thinking:* that television was so all-powerful that a perfectly good candidate could be ruined by one perfectly bad moment. Yes, Tim Russert was a media bellwether and the king of the Beltway.

But our supporters didn't work for the media. They didn't live inside the Beltway. They lived in Austin, Texas, and Seattle, Washington, and they knew better than to think that a television show *on Sunday morning,* watched by a fraction of the people in the country, could end this thing.

For that one moment, we were the ones, at campaign headquarters, who needed reassuring. And when we needed them, when we were falling, the people stepped up and caught us.

YOU HAVE THE POWER

"Today I announce that I am running for president of the United States of America." His trademark sleeves rolled up to his elbows, Howard Dean looked down at 5,000 people, a sea of swaying, blue Dean for America signs, on Church Street in sunny Burlington. "I speak not only for my candidacy. I speak for a new American century and a new generation of Americans—both young people and the young at heart. We seek the great restoration of

American values and the restoration of our nation's traditional purpose in the world. This is a campaign to unite and empower people everywhere."

It was Monday, June 23, 2003, the day after *Meet the Press* and the day Dean formally announced his candidacy. Gray Brooks, the clean-cut blond college freshman who had driven all the way to campaign headquarters from Alabama, introduced him with a speech that made you want to vote for Gray, too. And then Howard spoke with clarity and energy, and delivered the real message of the campaign:

> You have the power to rid Washington of the politics of money!
> You have the power to make right as important as might!
> You have the power to give Americans a reason to vote again!
> You have the power to restore our nation to fiscal sanity and bring jobs back to our people!
> You have the power to fulfill Harry Truman's dream and bring health insurance to every American!
> You have the power to give us a foreign policy consistent with American values again!
> You have the power to take back the Democratic Party!
> You have the power to take our country back!
> And **WE** have the power to take the White House back in 2004!

It was the largest campaign announcement in U.S. history. Because of the Internet, it wasn't just those 5,000 people in Burlington. We had 30,000 people linked up across the country, at about 400 events, as the web site proudly proclaimed, "ranging from half-a-dozen people watching the speech and eating cake at the Bodes General Store in rural Abiquiu, New Mexico, to more than 1,200 people packed in the San Francisco Hyatt Regency."

But once again, most reporters didn't get it, and the newspapers and TV networks only noted the crowd in Burlington as they blandly reported that after months of campaigning, another long shot candidate had officially joined the race—blah, blah, blah. To them, Dean had been run over and left for dead by Tim Russert.

But we were used to being underestimated, and the campaign had been reenergized by the amazing response of this shifting, moving, dynamic online base. And the money was pouring in—$200,000 on the day he announced. As soon as he finished his speech, the governor headed off on the road for a week of campaigning. Before he left, we met with him and told him that it looked good, that we'd easily hit our second quarter goal of $4.5 million.

Inside the campaign, we were always looking for ways to show our faith in the people out there, to involve them in what we were doing, to take our cues from them, to model the campaign on their passionate involvement. At one of these meetings with Stephanie Schriock, our young finance director, we tossed out the idea of posting our fund-raising—not just the results, like other campaigns, but the goal. Invite the people in and open up the books. Give them the knowledge and information—how much money we wanted to raise—and they'd take the responsibility for doing it.

We were in uncharted territory here. No campaign has ever announced the amount of money it hopes to raise. The reasons are easy to understand: First, you don't want the other campaigns to know what you're doing. This is like a football team sharing its playbook with the other team's defense.

But the biggest reason is that all-important question of momentum. If you tell the world that you're going to raise $4.5 million and you only raise $4.2 million—as big as that number is—you've just announced to the world that you're losing momentum. And in a presidential campaign, momentum is the *other* currency. One story about a candidate's failure to meet his fund-raising goals would have the sharks circling. No mainstream presidential candidate had ever made his fund-raising goals transparent for the world to see.

Which was all the more reason we needed to do it.

After the governor left to go out on the road, we talked about finding a symbol—something that could rise like a thermometer and show how close we were coming to our goal. But we didn't want to use a thermometer, like every hospital renovation fund and every church building fund. We needed something else, something easy to recognize . . .

"A baseball bat," said Larry Biddle, who was in charge of direct mailing in Stephanie Schriock's finance department. We all stared at each other. Okay. A baseball bat it was.

That's how, in the final days of fund-raising that June, a baseball bat went up in the corner of the Dean for America web site announcing our fund-raising goal of $4.5 million. And we just laid it out—presidential campaigns in America in the early twenty-first century were a race to raise money to buy TV ads. If we had any hope of defeating George W. Bush, it would be by raising enough money to combat his huge financial advantage. He'd already raised more than $100 million, mostly through his corporate and wealthy donors. If we were going to take the people's message to Washington, we needed to have the money to do it.

The next week was the most amazing thing I've ever seen in a campaign. All over the Internet, smaller, individual bats went up, as this huge network of grassroots organizations took it upon themselves to replicate the campaign at their own levels. All of the blogs that had followed the campaign began filling up their own bats, and the money began flowing in. We had hung white boards and butcher paper up all over the campaign headquarters to measure everything from Meetups to money, and the fund-raising chart—which still hangs in my office at home—tells an incredible story, broken down by the day and the hour and the half-hour and the dollar.

On Tuesday, $300,000 came in. Wednesday and Thursday: $300,000 each. And then, on Friday, June 27: $500,000. *A half-million in one day.* And this wasn't a bunch of wealthy Americans at a fancy dinner forking over a bunch of $2,000 checks between the chicken l'orange and the flambé. The average Dean donation was for about a hundred bucks. These were regular Americans—21,000 of them that one week—dipping into the grocery money to say that they wanted their country back.

In the early morning hours of Sunday, June 29, forty-five hours before the Federal Election Commission deadline, I got on the blog and wrote:

> As of last Sunday morning, June 22, the Dean for America campaign had raised $3.2 million in this quarter. Since that morning—beginning with the Sunday *Meet the Press* interview—we have experienced an unprecedented surge in contributions and have now crossed the $6 million mark. This is a surge in contributions of $2.8 million in just eight days. Of the $2.8 million over $2 million has come from Internet contributions . . .
>
> We wanted you to know what you had accomplished—that your hard work, and individual effort, when added up to the hard work and efforts of thousands of others, has made a huge difference for our campaign.

You proved that a rag tag team of a campaign and volunteers and self-organized grassroots can do what no one thought we could do—compete financially.

Now with less than 45 hours to go—we are able to set a goal that nobody would have thought possible even a week ago—the goal of reaching $6.5 million by midnight Monday June 30th—you have already proven the power of our numbers, and what can be achieved when each of us takes an individual action that is matched by the action of thousands of others.

At two o'clock in the morning on Sunday, we put up a new bat with the revised goal of $6.5 million. Five minutes after the bat went up, Governor Dean called Nicco from California, where it was almost 11 P.M. He was, understandably, freaked out.

"We've been hacked!" he told Stephanie Schriock.

"What do you mean we've been hacked?"

"Someone has gotten on our web site and there's a baseball bat that says we have a goal of raising $6.5 million!"

"But we do have a goal of raising $6.5 million," Stephanie told him.

The governor's end of the phone was quiet for a minute. "What the hell is going on over there?" When Governor Dean left, we had $3.2 million and a goal of $4.5. And, like all of the other campaigns, we kept our goal to ourselves. In the time since he'd left, we had not only upped the goal by $2 million, we had *posted the frickin' number on the Internet.*

"Campaigns don't do this," Howard said.

Stephanie, Nicco, and I explained the thinking of everyone from Finance to Field, from the Political desk to the web room—that we needed to take this thing all the way to *open-source,* put the code out there and see if the people could improve it, the way they had with posters and software and all the less significant aspects of this campaign. So why not also turn over some control of what was arguably the most important thing: Money.

"You know what?" Howard said. "You're absolutely right. Let's do this!" And then, as an afterthought, he said, "I hope this works."

It was a measure of how far the naturally cautious Dean had come in the last few months. After we hung up, Nicco and I also realized it was a measure of how far the self-described "technophobe" had come in embracing the Internet. This guy who didn't know a blog from a log a few months earlier was now on the road checking his web site on his lap top at

11:00 P.M., and calling in to tell us that we'd been hit by a hacker. We weren't sure if, when the campaign started, he would even have known what a hacker was.

On Monday, the money was cascading in. And we were drowning. So many people were clicking on Blog for America that it crashed and there was no hope of getting it back up that day. We were in the weird position of watching this thing we'd built for people to use break down precisely because so many people were using it. This was pretty much the Sisyphus-like task of Nicco and the other programmers; to build tools that the rest of us did our best to overwhelm and break. That Monday, the server that processed the online donations was the one tool we couldn't lose. But so many people were contributing that it was coughing and hiccupping and we have no idea how many donations we lost because people got tired of waiting for that little computer hourglass to tell them that their donation had gone through. All day and night Nicco was patching the thing with scotch tape and bubble gum. We were practically hugging the server. *Come on, baby. Come on. You can do it. Please don't die.*

At 10:00 P.M. Eastern Time, I got on the Blog and wrote: "Your efforts today were amazing and in my view historic. There is simply no precedent for this kind of response over the Internet."

But we had to keep fund-raising until midnight on the West Coast and so we worked in the web room all night and into the next morning, watching the numbers roll, until finally, at 3:00 A.M., Zephyr and I just stood there staring at the bat on Nicco's computer screen.

It said $7.2 million. "Holy shit," I said. We had raised $828,000 in one day.

No. *They* had raised $828,000 in one day.

"Uh, excuse me." There was a kid in the doorway of the web room, one of those Deanie babies who never seemed to need sleep. "I think you're gonna want to see this."

Dead on our feet, we followed him into the mailroom. During this entire week, while we were focused on what was happening on the Internet, bags and boxes of mail had been coming in from people who wanted to donate the old fashioned way—what techies derisively call "snail mail." Stephanie Schriock and Larry Biddle had put a great direct mail fund-raising campaign together with—as usual, little fanfare—and the result was a mountain of snail mail on the floor of the mailroom.

While we had focused on this new form of fund-raising, the older one had caught us off guard. The LLBSOS never failed to find a way to exceed our goals—Schriock had snowed me again.

In the end, there would be $400,000—almost none of it $2,000 checks, but fifty bucks here, twenty there, whatever real Americans could afford. In all, we had 59,000 supporters at that point, contributing an average of $112, a groundswell of average Americans that even the media couldn't ignore anymore. They swarmed our headquarters as the fund-raising story got out. Some reporters accused us of sandbagging—posting a low goal so that it looked amazing when we topped it. The only problem with that theory was this: It *was* amazing. Despite Schriock's nickname, none of us had any clue when we put a baseball bat up on a computer screen that it would translate into this. It was the most amazing thing I'd ever seen in politics. We had begun the quarter with the ambitious goal of getting to $4.5 million to try to catch the frontrunners, Kerry (who raised $5.1 million that quarter) and Edwards (who raised $5.8 million). Instead, we had blown past them.

We had taken the chance—and we'd talked Howard Dean into taking the chance—that if we jumped off that fifteen-story building, these incredible people would gather and catch us. What none of us had imagined was that when we finally nudged the campaign off the edge of the building, the damn thing would start to fly.

8

THE OPEN SOURCE CAMPAIGN

Hockey Sticks, Troll Bats, and the Sleepless Summer Tour

I WAS STARING at a hockey stick. It took me a minute to recognize it, but that's what it was.

In the past, when a business became profitable, its growth was assumed to be a slow, steady climb up a gradual hill—maybe at a rate of 5 or 10 percent a year. But in the 1990s, the technology boom was driven by Moore's Law, Gordon Moore's revolutionary observation that computer power would double every eighteen months, at the same rate that the size of chips shrank. With this exponential growth in capacity and speed, the vast amounts of investment capital, and the speculative nature of tech markets and products, the ascension of new companies was far more immediate and drastic (so too, it turned out, was their collapse). So a new company might slog along, break even for a couple of years, its line of profitability rising slowly or even staying flat. And then, suddenly, the company catches, the investment money flows in and the rise isn't gradual any more. The company hits a kind of tipping point, revenues take a sharp turn up, and they keep going up, not at 4 or 5 percent a year, but at a sixty-degree angle, the company doubling in size in a matter of months, sometimes for a period of years, like Microsoft or Intel. Step back from this growth chart and it looks like—a hockey stick.

I had been a small investor in a number of tech startups and worked for a handful of others (my love for underdogs doesn't end with politics) and we used to daydream about the moment we went hockey stick; we'd watch other companies enviously as their fortunes took that sudden turn up. But the hockey stick was something I had only seen on white boards in the conference rooms of tech companies, and in those exhausting and exhilarating final days of June 2003, as I stared at a graph of Dean for America fund-raising, it took me a minute to realize what I was seeing.

A hockey stick. On a presidential campaign.

We had been making solid progress since January, but then, in the last eight days of June, we brought in as much money as we had in the previous *eighty*. And this quantum growth wasn't just in our bank account. We saw similar leaps in the number of people organizing on Meetup.com and the number of hits to our web site, the number of links to Howard Dean blogs and sites and—once the media saw the amount of money we were raising, a concept they could grasp—the number of mainstream reporters making their way to Burlington.

It would've been nice to take time to celebrate, but honestly, we were all too tired. We'd been racing tirelessly toward this moment since January, when we sat in campaign headquarters and worked backward from the general election, staring at that white board with those impossible numbers that we needed to reach to challenge George W. Bush: two million supporters and 200 million dollars. Since that meeting, I'd been preaching that the Internet was the only tool capable of sparking the viral growth required to get there. I had driven the campaign staff, both online and off, with the idea that a presidential election is a marathon but that we were so far behind we needed to sprint the first five miles as if it were a hundred-yard dash. I'd decreed that as long as there was a chance the governor was campaigning *somewhere,* even on the West Cast, I wanted someone in every department at campaign headquarters until he was down for the evening—the much decried "Midnight Rule," which I religiously followed myself.

And so our overworked campaign staff, most of them young and inexperienced, had put in impossible hours and driven themselves to exhaustion to make it to the end of June, when the first five miles of the marathon would be finished. I didn't have the heart to tell them that we had twenty-two more miles to sprint.

The best thing about viral Internet growth is that you can double your size in a matter of days. The worst thing about viral Internet growth is that you can double your size in a matter of days. Suddenly we were among the frontrunners—just behind Kerry, Edwards, and Gephardt—yet we still had the tiny, inexperienced staff of a dark horse. We began hiring as fast as we could, but we could never catch up. If anything, our June fund-raising triumph only emphasized just how far behind we were.

No matter how successful we were at reaching our goals—everyone, including Howard Dean, understood how daunting our task was—at staff meetings at the end of each quarter the governor would start with the same talk "We got to the top of a mountain no one would have given us any chance of getting to—and guess what? Look at the one we have to climb this quarter. It's steeper and higher than the last one—and when we get to the top of it—the next one will be steeper still." Then he would thank everyone and urge us to work that much harder to keep growing.

All along, I had seen that first phase of the campaign—building the framework for a campaign that could grow from zero to $200 million in a year—as the toughest stretch. But once we built it, in some ways, the work was only beginning. Suddenly we faced problems no one had ever encountered before.

By summer, more than half our contributions were coming in over the Internet. We found ourselves in an era where our biggest worry wasn't John Kerry whacking us in a campaign spot. The attack we worried about could come from a fourteen-year-old kid in Thailand. Some guy in his garage in Oakland could take out a presidential campaign with nothing more than his $500 laptop.

Over the years, hackers had taken down all the Internet giants—from Amazon to Yahoo and everyone in between, media outlets, brokerage houses, retailers—with Denial of Service (DoS) attacks, jamming their web sites with self-replicating strings of messages, or using their computers to flood other networks with messages, so that someone trying to visit those web sites was denied service.

The Dean for America campaign was uniquely susceptible to a DoS attack. Our funding always came down to the last seven days of each quarter, when the momentum would build and supporters would inundate the web site with donations. After June, we quickly realized that if someone whacked us in the last week of the quarter, we were dead. We could lose hundreds of thousands of dollars, maybe millions over the course of the campaign.

We took several steps to be ready, including the drastic measure of having Nicco build a redundant system, a complete replica of our Internet server, in case the first system crashed. And in fact, we did suffer three

DoS attacks during the campaign (these were never made public for the same reason politicians don't make death threats public; you don't want to give lunatics any more bad ideas). The last attack was exactly what we feared, the big one, a full nuke coming at the worst possible time, on the last frenzied day of a fund-raising quarter.

But by that time we were ready.

By then, all the campaigns were embracing online fund-raising (though with nothing approaching our success) and both Kerry and Lieberman were using the same server infrastructure we were using, hosted by a company in Texas. So when the DoS hit, all three campaigns died. Lieberman's site displayed a message saying it was temporarily unable to process donations. Kerry's site didn't say anything. Both of them were down the rest of the day.

Our site was down for three minutes. Nicco immediately switched us over to our backup server and we hardly missed a beat.[1]

Another problem came from the sheer number and size of our donations. The Federal Election Commission had set up its reporting rules—requiring detailed reports fifteen days after the quarter ended—in response to the old political paradigm in which a candidate got seven thousand contributions of two thousand dollars each, spread out over a three-month period. These rules were designed to keep campaigns from hiding donors, from sneaking a Mafia boss in with the regular supporters. But no one had ever envisioned *99,000 contributions,* most of them for less than a hundred dollars, coming in the last six days of the quarter. This is one of the ways politics is going to have to change the way it does business, to let the average American back into the game.

At Dean for America, we nearly killed our poor compliance people. In an already short-staffed campaign, we had to put twenty people on FEC compliance alone, just to record every single $25 donation. When they were finished, after fifteen days, maybe an hour before deadline, the stacked report was fifteen feet tall.

Another welcome problem occurred with Meetup.com. Meetup was an amazing tool, but its founders hadn't designed it with politics in mind, especially a campaign like ours, and so over the summer, as we climbed toward 100,000 members, its limitations became clear. Meeting once a month

---

[1] The company that hosted our web site argued that the crash was caused by the tremendous amount of traffic to Dean for America. But if that were the case, it would follow that our redundant system would've crashed after we switched over, but it didn't go down.

was fine for knitting enthusiasts, but it was no way to run a presidential campaign. Sometimes people had conflicts on the first Wednesday of the month, or they wanted to leaflet on a different day. And as the numbers got bigger, most of the Meetup venues that could handle three or four hundred people turned out to be nightclubs. Yet many of our supporters were nineteen or twenty and couldn't get into bars. (And did we really want our strongest supporters getting all liquored up, anyway?)

So we developed GetLocal tools, not to replace Meetup, but to augment it. GetLocal was software that people could download onto their computers, which would let them enter a zip code and find the closest Dean meeting, or would help a Dean organizer find others in the area and get the word out to them to help clean up a park, or leaflet a neighborhood, or have a house party. For example, a single Dean activist in Los Angeles used GetLocal to find four people to staff a table at the Sunset Junction Street Fair, where they signed up five hundred thirty-eight more Dean supporters. By the end, there were untold thousands of GetLocal events for Howard Dean around the country, some social, others requiring labor-intensive organizing and community service. And the GetLocal tools could be used for all of them.

And in the best Open Source tradition, we put software out there for people not only to use, but also to improve, which they invariably did. Bloggers like Rick Klau were instrumental in designing and improving our software.

Sometimes, programs would simply show up over the transom: "Here, try this." Volunteers also stepped in to help our Internet people design the software for DeanLink, which was our version of the web site Friendster, which links people based on their mutual interests. DeanLink gave Dean supporters the chance to meet others like themselves. DeanLink also kept track of who signed the most people up to the campaign and the results showed the breadth of people drawn to the campaign. A forty-seven-year-old retired union organizer from Illinois brought in the second-highest number of people, but the highest was a fourteen-year-old computer buff from Sitka, Alaska, named Jonathan Kreiss-Tomkins, who signed up four hundred sixty-nine new Dean supporters. Jonathan, whose parents weren't even Dean supporters, visited campaign headquarters during his Christmas break. He flew from Sitka to Juneau to Seattle to New York to Burlington (making at least one leg of his journey on a cargo plane). He called his journey the Ideanarod.

One day, the phone rang at campaign headquarters and the man on the other end of the phone was lucky enough to get my wife, Kathy. His name was Lou Stark. He was eighty-nine years old and lived in Lake Elsinore, California. He wanted help printing out some Dean flyers to take down to his local library. He told Kathy that he'd been involved in politics earlier in his life. In a poignant series of phone conversations, he told her that in recent years he'd sort of given up on life, that he felt left behind by the world. That he thought God had forgotten him. So every day, he read the obituaries, to see whom he knew that had died, and waited for the day his own name would be in there. Then, one day, he heard Howard Dean on the radio and something clicked. He went out and brought a five-hundred-dollar personal computer so that he could go to the Dean web site and start reading the blog. Then he signed up for Meetup.com and soon he was the leader of his Meetup in Lake Elsinore. He said the campaign had given him a reason to live again, a reason to fight. Bob Rogan walked into my office after hearing about Lou from Kathy and just looked at me and said, "Joe, if this is the only good thing that happens in this campaign—the whole thing will be worth it." But it wasn't the only one—there were thousands of Lou Starks and they were all making a difference.

Every day we heard stories like this, like the elderly Wisconsin woman who died and, in lieu of flowers, instructed her family to have people donate money to a Democratic organization and her family chose the Dean campaign (according to them, the woman thought George W. Bush was "a lying whistle-ass").

Every day we met people or talked to people who had long ago given up on the democratic process, but were beginning to emerge into the sunlight again, like people who had been hiding out after a nuclear blast. I took to calling this incredible ongoing dialogue The Great American Conversation—a dynamic online discussion about the direction of our country. If our leaders weren't going to debate the war and the Patriot Act and other things, it didn't mean that Americans had to be silent.

And some amazing people began showing up at our doorstep to offer help, or just to see what was happening—experts on politics, government, and technology. Some of them were heroes of mine, people whose books and articles had been the building blocks for the campaign, without them knowing it, like William Greider, longtime *Rolling Stone* writer and author of such wonderful books as *The Soul of Capitalism,* and Richard Goodwin,

Even in 1957, I knew this machine would be the death of us. Pulling apart my mom's set. PEGGY SCHEUER

LASH TRIPPI '97

The wonderful Kathy Lash and I (with my kids Ted, Jim, and Christine, left to right) were married in 1997, in true pol style, with political buttons announcing the occasion. (Notice which one of us is the running mate.) NESHAN H. NALTCHAYAN/PAUL CULLEN (BUTTON)

Early in the Dean campaign we were so broke I went into a dunk tank to raise money—on my birthday. We raised $600 from staff for **Governor Dean.** KATHY LASH

My sons Ted (left) and Jim with Governor Dean and me aboard the Grassroots Express during the Sleepless Summer Tour, August 2003. KATHY LASH

The first stop of the Sleepless Summer Tour—the highlight of the Dean
for America movement—was Falls Church, Virginia. Four thousand people!

JOHN PETTITT/CLOUDVIEW.COM

I knew we were on to something amazing when the Sleepless Tour arrived in Seattle and found a crowd of fifteen thousand people waiting for us, August 2003. JOHN PETTITT/CLOUDVIEW.COM

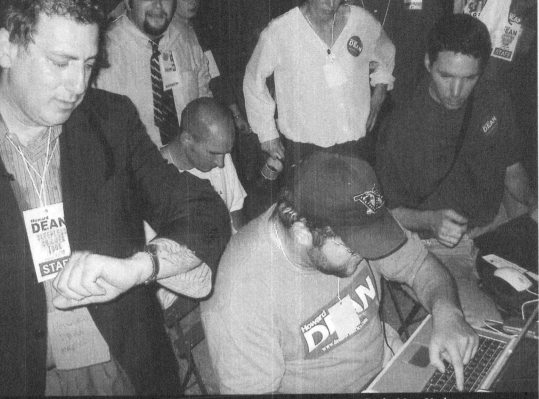

We gathered around the laptop buffet table at Bryant Park, New York, as the million-dollar deadline approached. From left to right: Me, Mike McGeary, Mat Gross (seated), Zephyr Teachout, Jim Brayton, and Nicco Mele, August 2003. GARRETT GRAFF

The giant screen in Bryant Park showed the web page and us breaking the bat, more than $1 million raised during the Sleepless Summer Tour, August 2003. VANESSA HRADSKY

The Dean for America staff celebrated the $7.6 million end of the second quarter of fund-raising, June 30, 2003. KATHY LASH

Kasey, the director of Canine Outreach for Dean for America, September 2003. JOHN PETTITT/CLOUDVIEW.COM

One of the few private moments between Kathy and me—as private as you can be when a *New York Times* photographer is in the elevator with you, October 2003. RUTH FREMSON, *NEW YORK TIMES*

Thousands of Dean supporters gathered for the Iowa Jefferson-Jackson rally in Des Moines, Iowa, November 2003. RYAN WITT

The day former vice president Al Gore endorsed Governor Dean was both the high point of the campaign and the day the other candidates began zeroing in on it, December 9, 2003 RYAN WITT

I used to tease the Internet team by asking, "Do you think John Kerry's staff is looking at the sunset?" December 2003. JIM BRAYTON

A gift from the Internet team: the photo of them watching the sunset, December 2003. JIM BRAYTON

Kathy at her desk in Burlington, Vermont, December 2003. JIM BRAYTON

Michael Silberman, Campaign Meetup director, was responsible for the success of the Dean Meetup effort in cities across the country, December 2003. JIM BRAYTON

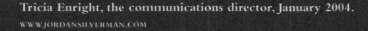

Tricia Enright, the communications director, January 2004.
WWW.JORDANSILVERMAN.COM

Political Director Paul Blank and I survey the crowd (and the snow) in the days before the Iowa caucuses, January 2004. RYAN WITT

After the election, Kathy, Kasey, and I returned to the Cummings Creek Farm on the eastern shore of Maryland to devote ourselves to more serious pursuits. ANDREW ROSSMEISSL

Nicco and Kasey crashing on the couch in my office, which always seemed to have an exhausted Dean staffer on it, July 2003.

JOHN PETTITT/CLOUDVIEW.COM

Just before Christmas, Kathy and I gave Howard a campaign fleece for the holidays, December 2003. JOHN PETTITT/CLOUDVIEW.COM

Howard and I congratulate each other after the endorsement by former vice president Al Gore. CALLIE SHELL

This is what happens when you raise almost $15 million in small political donations. Amelia Youngblood and Addy Gross needed a dolly to move the 12,000-page quarterly fund-raising report, September 2003.

JOHN PETTITT/CLOUDVIEW.COM

The Dean movement extended well beyond the border of a political campaign. Groups like this one organized themselves into Dean Corps volunteers. This group cleaned up a riverbank in Ithaca, New York, Summer 2003. LEE ANN VAN LEER

By the time we got to New Hampshire, the media scrutiny was overwhelming. (Left to right) Two unknown photographers, Dean, me, and Deputy Campaign Manager Bob Rogan, January 2004.

a former aide to Presidents John Kennedy and Lyndon Johnson, whose slender, transcendent 1992 book *Promises to Keep*—which never mentions the Internet—could have been the Bible for the Dean campaign:

> Our future depends on the ability to mount a struggle for extensive, even drastic changes in the institutions that compose both the private economy and the process of politics and government, along with the intricate web of relationships that connect them with each other and with the people. Changes so fundamental are revolutionary. But such revolutions pervade the entire chronicle of America. They have fueled our growth and progress while enabling us to sustain fidelity to the principles of American freedom.

It was thrilling to see the people drawn to this campaign. By the end of summer, there were so many Dean organizations, groups, web sites, and events out there no one could be expected to keep track of them all. All fifty states had vibrant Dean organizations and there were Generation Dean and Students for Dean groups on more than nine hundred college and high school campuses. There were Doctors for Dean and Kids4Dean and Mormons for Dean and every substrata of people you could imagine seemed to have a group or a web site or a chat room. And all of them wanted to *do something*.

At campaign headquarters, we honestly didn't know about many of the events going on out there. Unlike the classic Command and Control style of campaigning—where nothing happens that isn't scripted and tightly monitored by headquarters—if we wanted to know what was going on in the campaign, much of the time we had to do what everyone else did. We had to go where the campaign really was:

The blog.

BLOG FOR AMERICA

It was the nerve center of the campaign. The blogosphere was where we got ideas, feedback, support, money—everything a campaign needs to live. When the traditional media finally came around, beginning in late June, this was the hardest thing for them to grasp. They couldn't see these supporters and so reporters fixated on a bunch of people in basements, hunched over computers. But in fact, the blog was where the online campaign began its translation to the real world. And the first stop for people who wanted to get involved was often the campaign's official web log, Blog for America.

Like most web logs, Blog for America was divided into three columns. On the left was a list of links to other blogs, calendars, all the information people might need. On the right was the latest fund-raising bat, with our progress toward the goal marked in red.

Down the center was the main posting, often from our chief in-house blogger, Mat Gross, whose posts would be followed each day by hundreds of comments by readers and other bloggers. Most days, we'd post four or five times in this main section (and more toward the end of each fund-raising quarter) and so Mat also brought in all manner of guest bloggers: state organizers, authors, politicians, and regular supporters like the 100,000th Dean Meetup member, a woman from Denver who wrote about the first time she went to the Dean for America web site: "My sense of hope came back after reading the first few paragraphs—I finally found someone out there willing to work hard to *do* something about this mess."

Howard and I and the rest of the campaign staffers also delivered messages on the blog, and everyone quickly developed his or her own voice and style. Among the most popular and singular voices was that of Kate O'Connor, whose funny, folksy dispatches from the road were delivered in a tone made for blogging. Alongside the details of the latest rally, she would write about how Mary Smith had brought oatmeal cookies to a campaign event, but that the governor had devoured them all without leaving any for her, or how Bob Rogan had ripped his pants climbing out of an airplane. For me, she recalled David Haines—an unmistakably human, honest, noncondescending voice. She didn't talk down to people; it was like someone leaning over the booth in a restaurant to tell a neighbor something. This woman was the closest aide to a man running for president and yet people felt like they'd known her all their lives. Unlike corporate communications or the mechanized signature of candidates on most official campaign's correspondence, you knew there was a real person on the other end of Kate's blogs.

It was something I required of every campaign correspondence, that it be written by someone real, and that it be an authentic piece of communication. People are sick of getting a form letter from their congressman that starts "I wanted to personally inform you, **Mr. Joseph M. Trippi of**

St. Michael's, Maryland, about a key piece of legislation that blah, blah, blah . . ." These people are not morons. They *know* the letter was written by a junior staffer staring at a press release and that the blue signature at the bottom was stamped by a machine. The Internet is reversing the trend of corporate and political packaged communication and restoring old-fashioned *writing,* communicating to people in an authentic voice. You didn't have to agree, but when you read an e-mail or a letter or a blog from the Dean campaign, you knew there was a real person on the other end.

After eighteen years in politics, this was a harder lesson for the gover nor to grasp. His first blogs were notable mainly for their stiffness, along the lines of: *Thank you for your generous support.* At one point, Dean was asked to guest-blog for Lawrence Lessig, the visionary author and founder of Stanford Law School's Center for Internet and Society. It was a big honor, an even bigger opportunity to reach the movers and shakers in the blogging community, and an even bigger risk for someone who really didn't feel comfortable writing in a forum that is supposed to be equal parts honest, edgy, and informed. But one thing I admired most about Howard was the way he waded straight into deep water other candidates avoided. So he didn't have a clue how to blog—that was not going to stop him from becoming the first presidential candidate to do it.

In fact, at first the governor's posting was so flat and uninspired that some of the people commenting on Lessig's blog theorized that his comments had been ghostwritten by a nervous campaign staff eager to keep him from saying anything controversial (some even suggested it was done by a computer). So I got on the blog and immediately wrote, "I understand what you're saying, but if you really thought that these posts were being ghostwritten, don't you think we'd do a better job of it?"

David Weinberger wrote about this moment on the Joho blog, saying it was one of the most authentic political moments ever on the web, because when you looked at that exchange on the Lessig blog it was clear that this was *really* Howard Dean and it *really* was his campaign manager worrying about his blog. Readers of Lessig's site got to see a real behind-the-scenes moment, a candidate struggling to master a new form, much to the chagrin of his campaign manager.[2]

---

[2] We would eventually use several Internet experts and bloggers as consultants to the campaign, including Lessig, Weinberger, and Marcos Zuniga of Daily Kos.

So what makes a good blog? That's a tough one. The best answer might be the one that former Supreme Court justice Potter Stewart gave forty years ago when he admitted that he couldn't necessarily define pornography, "but I know it when see it."

Quoted in a *Wired* magazine story about the Dean campaign, Lessig said that the key to a successful blog is "to turn the audience into the speaker. A well-structured blog inspires both reading and writing. And by getting the audience to type, candidates get the audience committed. Engagement replaces reception, which in turn leads to real space action."

The best blogs feature a strong, *human* voice and a vibrant, running commentary by readers—instant response and feedback to what they are reading. This feedback was the thing missing from our first blog, Call to Action, but it became the centerpiece of Blog for America. We went from a handful of comments to a hundred to a thousand comments a day on the official blog, instant reaction from the people we were trying to reach. Also key to a good blog is a blog roll, a list of other recommended blogs. This creates a web of interconnected people debating serious issues, a running forum of political and social debate ranging all over the web.

These blogs ran from the deeply personal to the deeply political, but most shared the assumption that you aren't getting the whole story from traditional media, because of some combination of subtle (or not so subtle) bias, intellectual sloppiness, paternal arrogance, car-wreck sensationalism, edge-dulling centrism, and—maybe worst of all—conventional wisdom groupthink. And so people with some expertise in a certain area—or just an informed opinion—would get online to offer the other side to a big news story, or to point out what the news media had missed, or just to offer their two cents.

There has been some hand wringing over these bottom-up pundits and amateur news gatherers. A cynic once compared it to going to a restaurant and having the guy at the table next to you cook the meal. It's an amusing image that totally misses the point. The amazing thing is that people are doing this at all, committing themselves to debating and covering serious issues. The bloggers would also argue that in some ways, they are more accurate than traditional media, because if they make a mistake, five people immediately remark on it. They don't have the luxury to tuck a correction onto the Letters to the Editor page three days later. Traditional media are opaque; the blogosphere is nothing if not transparent. Unlike some reporter for the *New York Times* or the *New York Post,* a blogger's opinions and biases are right out front for everyone to read. Hell, a blogger *is* his opinions and biases—that's the whole point. And the final argument for blogger trust and veracity is that most bloggers are doing it *for free.* Bloggers recognize that political bias is less insidious in modern media than economic bias—in which news is just another kind of marketing, as packaged and focus-group tested as politics and music and dish soap.

An average day in the Dean blogosphere reflected all the energy of the campaign. On Blog for America, Mat might write about a news story critical of the governor, and hundreds of people would post comments about it, everything from deeply researched rebuttals to "Hang in there!" to discomfortingly plausible conspiracy theories about the media, or responses from the other candidates. Kate might weigh in from the road with pictures of the governor eating apple pie at a campaign event, and fifty people would post comments, everything from "Save me a slice!" to recipes for sour cream-raisin-green apple pie. People would randomly offer strategic advice, improvements on software and—uh, poetry:

> We got the boss that's keen
> His name is Howard Dean
> He's sharp as a tack
> And his record's clean—We thought Bush was fine
> But then he crossed the line
> 'Cause power gets you drunker
> Than Ripple wine.

As the daily traffic on Blog for America increased and the number of comments rose—from a few dozen to a few hundred to a few thousand—we used to celebrate every time they overwhelmed the damn thing and crashed the blog. The Web Team would work their asses off building it so that it couldn't be broken, then high-five and toast each other when it crashed, and then set about fixing it again so that it couldn't be broken. When it came back up, the people out there would be in awe of themselves: *We crashed it!*

But the official blog was only one voice among thousands.

Dean was the also subject of blogs by his supporters, by issue-driven activists commenting on all the campaigns, and by politically savvy, objective bloggers serving as de facto campaign reporters. Every day, the campaign was being written about—and sometimes second-guessed—by independent blogs like Daily Kos (written by the canny Marcos Moulitsas Zuniga, who was initially a Wesley Clark supporter), Atrios, and Talking Points Memo (Joshua Micah Marshall)—writers every bit as talented as anyone working in the mainstream media, but without the mainstream media's addiction to stale conventional wisdom, or its restraints, limitations, and time constraints, and with daily readerships that some newspapers would've killed for. We took such blogs seriously, listened to their advice and had Governor Dean do e-mail interviews with them when they asked. Even when the networks finally came around, we remained loyal to the blogs, who were writing about us back when Peter Jennings couldn't bring himself to mention Howard Dean.

The bulk of the daily blogging about the Dean campaign, like the campaign itself, came from grassroots organizers. People just doing their own stuff. And the stuff they came up with was amazing. We didn't just monitor these blogs. We listened. We took the feedback seriously. There were times that we altered Dean's stump speech based on a suggestion from a blog, or improved our campaign strategy. I can't count the number of times we turned to the blog for help.

One of the most difficult assignments for any presidential campaign in a primary is to make sure the candidate qualifies for the ballot in the primary and caucus states. This convoluted process is different in each state and is a huge undertaking, requiring local expertise, organization, and manpower. Often, smaller campaigns are left out in the cold. Even the big frontrunners can have trouble with this; John Kerry's campaign failed to get enough signatures to qualify a full slate of Kerry delegates in the key state of New York. We turned it over to our Internet folks, who did much of the heavy lifting to get us on the ballot there (a formality that, as it turned out, we wouldn't need).

It became obvious pretty quickly that a couple of dozen sleep-deprived political junkies in our corner offices of the Dean campaign couldn't possibly match the brainpower and resourcefulness of six hundred thousand Americans. We couldn't see every hole and every flaw that they could see. For one thing, we were too close to it. There were any number of times one of those six hundred thousand came up with something that should have been obvious to us. There were countless instances where we were trying to figure out our next move and someone on the Internet came to our rescue.

One problem with running a campaign open source is that you also open it up to every crank with a computer who despises your candidate. We had a policy that we wouldn't remove critical posts from the blog. If people wanted to take an honest shot at Howard or at the campaign, we welcomed it. Disagree with his stance on the war? Fine. Want to argue abortion? Great. It sparked some of the most interesting debates on the blog to have differing opinions. Liberal Dean supporters could go on for days within themselves over his stance on the death penalty (he was guardedly for it) or gun control (he was guardedly against it). But there were a few people who had no interest in the issues; they only wanted to mess with the blog. One guy would cut and paste the words Dean Sucks and then drop 400 pages of this shit on the blog, Dean Sucks over and over. It would go on for pages, a long string of Dean Sucks, so that no one else could post.

Bloggers call them trolls—those anonymous Internet vandals whose only goal is to sabotage web sites and blogs. Journalists and people from other campaigns were often convinced that our reliance on the Internet left us more open to trolls, pranksters, and dirty political tricksters. This is the rationale that corporations and campaigns sometimes make against the transparency of an Open Source Internet presence. But we argued that the same people who tried to crash our web site could also tear down a Dean sign from a building; the same people who misrepresented themselves on-line as our supporters could put on a Dean shirt and misrepresent us out in the real world. There is no technology that trumps human nature—either its best attributes or its worst. In the campaign web room, we removed the Dean Sucks troll from Blog for America every time we saw him, but the other bloggers came up with the most ingenious solution for the problem. They formed into Dean Teams who would—among other duties—post what they called a Troll Bat, another fund-raising bat, so that every time the Dean Sucks guy or some other jackass tried to disrupt the site, Dean supporters would spend an hour donating in this guy's name ("Feed the goal, not the troll!") and let the troll know that he was personally responsible for Dean getting another $10,950 in campaign money: "A big thank you to our troll sponsor for making this all possible." In many ways, the blog was self-policing and our Internet community was an amazingly civil place. And an extraordinarily creative one.

In July, the media reported that Dick Cheney was having a $2,000-a-plate lunch to raise money from Bush supporters. It was a blogger who came up with the idea to put up a Cheney bat that day, along with a live streaming web-cast of Howard eating a three-dollar turkey sandwich. We called it the Cheney Challenge and it was the kind of idea that reinforced what the campaign was about. Bush and Cheney were courting those wealthy $2,000 donors, the richest of the rich—one-tenth of 1percent of the people—and we wanted to represent the rest, the ones who could only afford $25 or $50, the ones who were tired of a system that seemed designed to ignore them. And we found people on the blog committed to beating that system. With his 125 or so lunch donors, Cheney raised $250,000 that day. Our Cheney bat raised more than $500,000 from 9,700 donors giving about fifty bucks each. And we didn't think of that. *They* thought of that.

It was also bloggers who organized Dean Corps—a sort of low-impact, weekend Peace Corps. Because they could organize themselves, and didn't have to wait for someone on high to tell them what to do, these people took the energy from Dean for America and spread it around their communities, in areas that had nothing to do with the election.

They got together in neighborhoods to clean up riverbanks, to read to children, and to collect food for homeless people. People were cleaning creeks for God's sake, just putting Creek Clean Up on the Net and waiting to see who showed up. This wasn't for the campaign; this was for their community. *We* weren't doing that. *They* were doing that.

Weinberger, the author of the seminal book on the subject, *Small Pieces Loosely Joined,* told *Wired* magazine's Gary Wolf that the structure of the Dean for America campaign, what Internet theorists call "end-to-end" architecture, was the key to its success:

> The goal is not necessarily to have messages flowing up and down.
> Democracy is supposed to be about people talking with each other about
> what matters to them—Ideas from the grass roots don't have to go back
> up to headquarters to be adopted—The Dean campaign instead gives
> you the tools to institute your ideas without involving headquarters.

That was the beauty of the Dean campaign in the summer of 2003, the fact that it *wasn't* the Dean campaign. And so, even as the media finally was swinging its massive attention around to covering Howard Dean's growing momentum, these same networks, newsmagazines, and daily papers were continuing to miss the bigger story.

People all across the country were engaging in civic life again. Their faith in America and in their own power was being restored. Howard Dean was doing something more important than running for president; he was creating a political movement in which the people mattered again. That was the real story. And it was just beginning.

THE OPEN SOURCE CAMPAIGN

On those rare occasions that I got any sleep, I'd go to bed wondering: Is this as far as you can go? Is this as *open-source* as you can take a political campaign in the year 2003? Isn't there anything else you could do? We'd talk about it in meetings, what else we could do—other ways we could show faith in the people out there, other 15-story buildings that we could jump from.

By summer, just about everyone inside the campaign was a believer in the concept (if not always the ramifications) of an open source campaign, from the governor on down. Even those people most dubious in the beginning about using the Internet to grow the campaign could see that something amazing was happening, something bigger than poll numbers and fund-raising.

When reporters finally came around to ask about the Internet, I had to tell them: this is not just about the Internet anymore. Our first challenge had been to take the campaign—a traditionally offline construct—online. We had done that. Now we had to take it back offline. Already, there were countless examples of this happening—some of them happening to me. One day, I was standing in line at an airport being searched by one of those serious, unsmiling Transportation Safety Agency people when another TSA guy leaned over and said under his breath, "Hey, I get your e-mails." Then, while I did a double-take, he straightened back up as if nothing had happened. Other times, I'd be at campaign events and hear someone call my name and I'd look up at an unfamiliar face. "Hey, it's me. Rick Klau!" And we'd hug like two people who'd gone to summer camp together, even though we'd never met. Other bloggers like CD Marine would post pictures of us from a campaign event then would be shocked that I remembered them when we met again months later.

"Are you kidding me?" I asked. "CD, you're one of my favorite posters on Blog for America. Of course I know who you are." They couldn't believe the campaign manager was actually reading their messages and ideas and knew them. But there were people who were only screen names to me that I grew to care about, like CD Marine, and Patty from Vermont, the amazing mom of two energetic sons, who worked her heart out for Howard Dean.

And it wasn't just me. Mat, Zephyr, and Nicco, Karl, and Tricia, everyone had stories of making connections with people who had only existed as e-mail addresses to them when the campaign started. It was ironic. Right about the time this thing was moving from theoretical to practical, the media was just coming around to cover the theoretical part.

Together with our grassroots partners, we had built from scratch a machine capable of flying. And it was only natural that we'd want to take it out for a spin.

THE SLEEPLESS SUMMER TOUR

The name was Tricia Enright's, and it was perfect. The organizing was done by Tamara Pogue, and it was amazing.

In August, we decided to take the campaign *out there* and show the people what they've built, show the world what they've built. Load up a cramped chartered jet with the governor and two dozen staff people—all those voices on the blog that people had gotten to know—invite twenty-five or so campaign reporters and see what happens. Ten cities in four days at the end of August, from one end of the country to the other and back again—a vibrant, once-in-a-lifetime display of the power of all these people reconnecting with democracy.

Our goal was to raise a million dollars over those four days, but more important, we wanted to demonstrate the strength of this thing we had built. This wasn't a campaign sending out advance people to build a crowd of supporters. This was a campaign doing nothing more than building ten stages in ten cities and trusting that the people on the ground would carry us the rest of the way. This was a campaign venturing out at the worst possible time for politicians, the dreaded dog days of August, when campaigns traditionally figure that people are too busy at the lake or at their kids' baseball games to listen to a candidate drone on about Medicare, when the usual conventional wisdom says to just throw up your hands, take a few weeks off, and pick it up again in the fall. If you needed to give away free breakfast to get five hundred people to attend a political event in April, in August, you probably had to give cash money to draw more than a couple dozen. Yet here we were trotting out the whole campaign on little more than a hunch, betting that the people would show up. *And* we were bringing along the skeptical national press to witness what could very well be our abject failure, the sudden end of the campaign in a flurry of hyped expectations.

Personally, I couldn't wait.

We'd just gotten the covers of *Time* and *Newsweek,* but the real measure of our campaign was out there. For me, this was the final step in us turning the campaign over to the people. Forget jumping off a building. I was ready to jump from a chartered plane.

Throughout the campaign, Kathy and I lived in a small, secluded mother-in-law apartment in the back of a house in Burlington. On August 19, I took my first "day off" since January, and Kathy and I raced home to the eastern shore of Maryland, where I slept in my own bed for the first time in five months. In was nice to see the place, even if we had to be careful not to disturb the two Carolina wrens that had built a nest on top of our china cabinet and hatched three chicks while we were gone. I don't know who was happier to be home, me, Kathy, or our terrier Kasey, whose picture on the web site identified her as the campaign's "Director of Canine Outreach," and who raced around chasing birds and cats and randomly throwing herself into the pond.

The next day, we had a Dean Ice Cream Social at the house and then we drove to Falls Church, Virginia, to join up with what we called the "Grassroots Express," for the $1 million Beat Bush Challenge and the first leg of the Sleepless Summer Tour. From that first rally, this was not like any campaign event I have ever seen.

There were four thousand people at that first event in Virginia, waving their arms and screaming—it felt more like a rock concert than a campaign event. We sardine-packed onto the plane ("with legroom sufficient only for the legless," the *New Yorker's* Mark Singer wrote) flew from Virginia to Milwaukee and another eight hundred people and then a runway stop at the Boise, Idaho, airport. Everywhere we went, signs proclaimed: "People-Powered Howard!" and "Take Back the Country!" and the crowds went crazy over Howard's stump speech, even as his voice became a dry husk. In

Boise, he grabbed an American flag and said, "You see this flag? This flag does not belong to John Ashcroft and the right wing of the Republican Party. This flag belongs to the people of the United States of America! And we're gonna take it back!" In Portland, Oregon, a crowd of three thousand gathered in early afternoon, on the street and on neighboring rooftops, swaying and yelling for a political candidate, a *political candidate,* more than a year before the election. All of it captured by John Pettitt, a successful, gifted Internet entrepreneur who was drawn to the Dean campaign as a volunteer and soon was indulging his other love—photography—as our campaign photographer.

It was amazing to go to these rallies and see huge signs whose slogan—"You gotta believe!"—I'd had to explain to the young Dean staffers only months earlier. But as great as those early stops were, it was when we landed in Seattle that I knew that this thing was huge, that it was bigger than even I had imagined.

As the *Seattle Weekly* later reported, the event there was planned with "little advance buzz or advertising," nothing more than a stage thrown up in front of the Westlake Mall for a little afternoon rally. We pulled up and I wasn't sure I was seeing right. I climbed up on the stage and my jaw dropped. The crowd filled the courtyard and just kept going—*more than ten thousand people*—around trees and across streets, blue Dean signs as far as the eye could see, an army of people of every age and income and ethnicity, chanting the campaign's *real* theme: "I AM Howard Dean!" In what may be the most unbelievable thing that happened in the campaign, I was at a loss for words. (Other than "holy" and "shit.")

Then we stopped for breather at a community center in nearby Spokane, in front of an expected crowd of two hundred people. The governor began speaking to a crowd outside that looked slightly bigger than we'd expected. No, someone explained. This was the overflow, one of *three* overflow areas. In all, about one thousand people had gathered, five times what we'd been told to expect.

The crowds didn't just clap at the appointed applause lines. They yelled and sang and chanted like they were at a football game—*You say Howard! I say Dean! Howard! Dean! Howard! Dean!*—or raised their hands and chanted along like believers in an evangelist's tent.

For four days, we lived on that plane, eating our meals on the little fold-down trays. We alternated between elation and exhaustion, falling asleep for a few minutes on each other's shoulders or with our faces smashed against the oval airplane windows, jolting awake when we landed in a new city. The third day was Texas, a rowdy three thousand in San Antonio and five thousand more in Austin, right in George Bush's backyard. "Let's send George Bush back to where he's from—Connecticut!" The energy was amazing.

One of my biggest fears in taking another presidential campaign was that I would never get to see my kids, who live in Illinois with their mom. At one point we were being ushered along the rope line—people reaching out to shake hands with the governor, or just to touch him—when I heard "Dad! Dad!" I looked up to see my seventeen-year-old daughter, Christine, making her way through the crowd. She had flown up to be part of the campaign. My sons Ted, who was twelve, and Jim, fourteen, even came for the Sleepless Tour. Jim engaged in a sign war with Lyndon LaRouche supporters in Seattle and volunteered to work sign-up tables.

In Austin, I got separated from Ted, who wants to be a documentary filmmaker when he grows up. We ended up on separate press buses, and he arrived at the event first. I'll never forget the feeling as the governor, Kathy, and I walked into the hall. The room broke into a huge roar of applause. And for some reason, I looked up at the press riser. There was Ted, a set of huge headphones sitting cockeyed on his head, running a jittering network camera and grinning from ear to ear.[3] I could feel the emotion bubbling up, an elation so deep you're not sure it is going to come out as laughter or tears as this kid who had nearly died just a few years earlier sat smiling and waving at his dad from behind a $20,000 network camera. I hadn't worked on a presidential campaign since 1988, when Christine was a baby and before my sons were born. To have them beside me for this one, to have Kathy there with me, at the triumphant moment of the campaign of my life—it was almost more than I could bear.

But Kathy and the kids weren't my only family on the Sleepless Summer Tour. I couldn't have imagined how I could grow closer to these people I'd worked alongside for the last seven months—Zephyr, Nicco, Tamara, Mat, Drymala, Tricia, Paul, CarlwithaK, and so many more—but I did. These

---

[3] When I left the campaign, all the television news channels and networks were trying to score the first interview with me. I called the photojournalist, Felix Schein of NBC, who had given Ted his camera and told him I'd do whatever show he wanted me to do.

people had worked as hard as anyone I'd ever seen, young men and women who lived up to every oversize legend from the tribe of pols that I'd so eagerly joined twenty-four years earlier on the Kennedy campaign.

The last day, August 26, we flew from Chicago to New York. On the tarmac of LaGuardia Airport, with the sun going down around us, we huddled with the crew, like a football team of rumpled, tired, walking dead.

"Remember," I said to them, the words catching in my throat, "no matter what happens—nobody can ever take this away from you." We cried and hugged and I yelled, "Now back to work, damn it!"

For four days, the money had been pouring in on the web site: *$700,000* by San Antonio, and on the web site the bat was filling with red, but a million dollars was a huge number. *$800,000.* Some Bloggers wrote that they were worried that we wouldn't make it, that we'd set the bar too high. *$850,000.* There was just one day left! It would take an amazing push. Smaller bats went up all over the Internet. *$900,000.*

The event that night was a huge rally at Bryant Park in Manhattan. The press was furious with us because we were late getting into New York and they hadn't eaten. A candidate could theoretically survive a sex scandal, illegitimate children, drug usage, draft dodging, *and* membership in the Communist Party, but if you don't feed the reporters every five hours, your campaign is over. So we stopped the press bus at a little deli and everyone went in to get something to eat. That's when someone on the web team in Burlington called me and said, "Joe there's this cool idea on the blog. If we can somehow get to a million dollars out there before he goes on stage, they want him to carry a red bat on stage and say, 'You did it!'" I thought it was brilliant. It would mean very little to the reporters and pundits, but those thousands of our supporters watching on C-SPAN, or watching streaming feeds on their computer screens would know the significance, the bat filled with red on the web site and a red bat in the hands of the candidate. It was perfect. Months earlier, we had jumped and the people had caught us, and now this was the real test of their ownership in the campaign, that someone could somehow suggest it on the Internet and then, forty-five minutes later, see us do it.

I turned to Matt Vogel, the poor staffer who drew the short straw and happened to be standing next to me. "Go get a red bat."

He stared at me. "What?"

"A red bat. Go find a red bat."

It was after eight o'clock at night in New York City. This poor kid stared out at the busy street outside the deli. "But where—where am I supposed to get—"

"I don't care where you get it. Just go get a frickin' red bat."

Matt stepped out of the deli, looked to his left, paused, looked to his right, and wandered away. We weren't sure we'd ever see him again.

Bryant Park was electric, twelve thousand people locking arms and waving and chanting *Dean, Dean, Dean* while the singer JoAnna wailed out our theme song, "We Can," and my irrepressible wife stood at the front of this huge crowd bouncing and singing to the music. We had decided to go live on a wireless feed from Bryant Park to the web site, so Nicco, Zephyr, and Mat huddled backstage over a buffet table of laptops, grinning, smacking the refresh button, watching the online bat fill with red. $950,000—$960,000. The traffic on our web site was overwhelming. Eventually the site crashed. The blog crashed and our 800 number crashed. Finally, we overloaded the wireless network in Bryant Park and we had to send someone across the street to a Kinko's to update the bat. We had one person in Kinko's, someone else calling from Burlington, and Nicco in the park, constantly relaying numbers back and forth, so we'd get the right numbers on the giant board in the park. In the meantime, the crowd was screaming for Howard. He was supposed to go on at 10:00 P.M. and at quarter 'til we were twenty-five thousand dollars short—and still no bat. I was prepping the governor for his speech, telling him that when he got the bat, assuming there was a bat, he shouldn't wave it menacingly, but should just hold it up. He practiced holding an imaginary bat and—

"Trippi!" Nicco screamed at me. He had the phone to his ear. I ran over, looked over his shoulder and saw that he was filling the baseball bat. A huge number stared at me from the screen: $1,003,620. I checked my watch. 10:00. I couldn't believe it. We were high-fiving and hugging and just then someone on stage called out, "Ladies and gentlemen, the next president of the United States—Howard Dean!" And that's when, in my peripheral vision, I saw this kid—a red bat in his hand—running down the sidewalk like a lunatic, people stepping out of his way, and just as the governor was hitting the top of the stage, the kid threw the bat up, the governor caught it with one hand, walked up on stage and said, "You did it!"

It was a moment I'll never forget. A late summer night in New York City—the energy of that crowd—the energy of all those people who had willed us across the country—Howard on stage: *You have the power to make this party stand for something again, you have the power to take this country back. It's your power!*

I remember standing near that stage, looking out at those twelve thousand waving, cheering people and thinking that this was something totally unexpected. These people weren't celebrating how much money we'd raised, or Howard's picture on the cover of *Time,* or the polls that showed us taking the lead, or how hard we were working in the campaign, or how smart the kids in the web room were. This wasn't even about the governor up on that stage. These people were yelling for themselves. This was the first campaign owned and operated by the American people and they were letting out a primal, exultant cry that this thing, this system, this country, it was theirs, goddamn it, and they were ready to take it back!

I sometimes wish the campaign could've ended at that moment, with me standing in the shadows of that stage, the governor waving a red bat to a joyous crowd, the dark, twisted giant of traditional politics still off sleeping somewhere.

9

THE FALL

A .357 Magnum, Al Gore, and the Left Wing Freak Show

T HE TROUBLE STARTED in the early morning hours of July 1, a few
hours after our June fund-raising triumph. We were totally destroyed—Nicco, Zephyr, and me. We staggered out of the office to get
some sleep before the sun came up. I was standing in line at a deli to get
something to eat and the next thought that passed through my brain had a
startling clarity: Wow, I guess you *can* fall asleep on your feet.

I'd always thought that phrase was an exaggeration until that moment,
when I fell asleep, toppled over onto the deli case, and cracked one of my
ribs. Although I didn't know it when I came to. For a couple of weeks, I
just plugged along. Luckily it only hurt when I (1) moved or (2) breathed.

The governor, who sometimes has the physicality of an old jock, liked
to squeeze my side and playfully offered to diagnose me. "Does that hurt?"

"Yeaaugh!"

This went on for a few days until the governor seemed to realize that
I was actually in pain—stopping a senior staff meeting at one point, dead
in its tracks, to ask what color my stool was.

You have no idea how disconcerting it is to have your candidate for
president of the United States of America suddenly go doctor on you and
ask—well, questions a doctor would ask. That was enough for me.

I asked Kathy to get me an appointment with a practicing physician.
So she called the other Doctor Dean.

I drove to her office, in an old creamery near Shelburne, a few miles from Burlington. It was a Sunday and she walked around the office, switching on lights and small talking. She was thin, with glasses and dark hair to her shoulders, and, although a bit reserved, one of the nicest people you'll ever meet. Dr. Judith Steinberg Dean, my boss's wife, calmly informed me that I had broken a rib. Of course, there's nothing you can do for a broken rib except lie down for eight weeks and wait for it to heal, which was a luxury we both agreed I didn't have.

There's a baseball movie that came out a few years ago called *For the Love of the Game*, starring Kevin Costner—one of those movies that women think is corny and men think is moving. It was about an old pitcher playing in his last year for a losing team in a meaningless game, who goes out and inexplicably throws the game of his career—with his forty-something-year-old arm about to fall right off his shoulder—a perfect game the last time out. I'd let a few people around the office know that this was how I felt, like the old patched-together pitcher in *For the Love of the Game* and they'd taken to calling it "Trippi's movie." Nicco and I would greet each other with our version of the signature line from that movie: "We suck. But right now, we're the best team in baseball."

It became a campaign motto and the old baseball player's ethic became shorthand for the way I approached the physical demands of a presidential campaign. *Patch me up, doc. And send me back in there.*

In that spirit, I asked Judy if she could prescribe a painkiller. She said no. They were addictive.

I was in no mood for some Vermont naturopathy lecture. "Right," I said. "I hear they're also good for *pain*."

She just laughed. "Sorry, Joe."

Of course, it was well known around the campaign and in the media that Howard's wife wanted almost no part of this running for president business. She was a serious, well-respected physician and I admired the fact that Judy didn't feel the need to play the doting first-lady part just because everyone else said she should. But it made for some tough times on the campaign trail. We constantly had to explain her absence, and even when we explained it, reporters looked at us cockeyed, as if there had to be more to it than that. No, we said, she has her own career and while she supports Howard, she's not going to be taking an active role in campaigning. Having the candidate's spouse do some press could have taken the pressure off him and bought a little more needed publicity, but Howard was adamant that this campaign would not get in the way of his family or his respect for his wife's career. The Deans were simply not going to play the game just because everyone else did. The ironic (and sad) thing was that here was the most loving, real family I'd ever seen in politics—behaving the way people should behave—and the press wanted to know what was *wrong with them*. Reporters were so used to candidates' Stepford families and that packaged, posed campaign domesticity that they missed the real thing when it was right in front of them.

While she was looking at my ribs that day, Judy asked if I was having any other health issues. Before 2003, I had gotten into a pretty good rhythm caring for my diabetes. During the off years, when I didn't work on campaigns, I got plenty of rest, ate right, ran six miles a week, and generally kept my blood sugar under control. Then, during campaign season I would go all out again, work myself sick and watch my blood sugar go through the roof. With diabetes, you feel it in your nerve endings— your fingers and toes first, a kind of burning that is unmistakable. During campaigns, my fingers and toes would feel like they were on fire, and it wouldn't go away until weeks after the campaign ended.

"Well," I said, "I have this burning in my fingers and toes."

She stared at me. Like every other doctor when they hear that symptom, she knew exactly what I was saying. "You have diabetes."

"Yes," I said.

"I see." She knew I hadn't told Howard. We talked a little bit about the demands of a campaign like this—fatigue, bad diet, lack of exercise, and most of all, stress—practically the recipe for exacerbating diabetes. She scrounged up some samples of Glucontrol, a blood sugar medication that I'd been taking for years, and gently chastised me for not keeping up with it. From that point on, Judith Steinberg Dean was my doctor on the campaign. But to the best of my knowledge, she never shared a single detail about my condition with her husband. I remember him talking to her on the phone once, half-joking and half-seriously frustrated that she wouldn't give him the details of my exam. "But I'm your husband. And I'm his boss—What do you mean, doctor-client privilege?"

When I took the Dean campaign, I was still wasted from the Holden congressional race in 2002. This time, I hadn't taken my usual year to recharge. By late summer, I could tell that something was wrong. I had never felt this bad in my life. Apparently, other people could see it, too.[1] Staffers urged me to take some time off, but an insurgent presidential campaign isn't something you take time off from, especially one that is running twenty-four hours a day on the Internet. "When am I supposed to take time off?" I would ask my sweet, nurturing assistant, Kristen Morgante. "Suit yourself, asshole," she'd say. "Die. See if any of us gives a shit." The truth is that Kristen probably did more to keep me alive during the campaign than anyone except Kathy. I hope she knows how grateful I am, and that she doesn't regret it.

After the Sleepless Summer Tour, we sensed our moment to grab even more momentum and to frame the agenda, and so we spent about a million dollars on television ads in states like Washington, Michigan, Arizona, and Texas, assuming we would raise several times that by attracting new people to the campaign and the web site. As it turned out, we raised about $1.4 million, enough to pay for the ads, certainly, but some people criticized us for "going up" too soon, wasting money on ads in places where the primary was still weeks or months off. Obviously, the closer a spot comes out to the election, the more effective it is. But this was the old paradigm of campaigns, the chicken dinner theory of raising money to buy ads to go on TV. We were turning it on its ear, going on TV to get people to go online to raise money. We saw a chance to seize the debate, so we took it.

What we didn't see was Wesley Clark. The former general was considering jumping in the race and in September, I went to Los Angeles with Howard for a stressful (and ultimately ill-fated) meeting to feel out Clark. Of course, not long after our meeting, Clark *did* jump in the race, the first real blow to our campaign after seven months of mostly easy sailing. Suddenly, we weren't the only Washington outsider in the field anymore. And the *other* Washington outsider was a goddamned Rhodes Scholar war hero, hand-picked by Bill Clinton. Instantly, Wes Clark became the one guy we most feared. And now he was in the race.

After the Clark meeting, I flew on a small Cessna up to Northern California, to the home of Steven and Mary Swig, who had graciously agreed

---

[1] In a profile of me in GQ magazine, Lisa DePaolo wrote: "He's ill. At times it is painfully obvious. All of which certainly raises the question: Is Howard Dean worth killing yourself for?" My response was that if GQ was doing a story on *me*, then we had to be on the verge of the apocalypse anyway.

to host a gathering of major donors like themselves; we called it the Dean's List Retreat. The Swigs had a beautiful home in the Redwoods and I stayed up late that night, talking to supporters and giving them an update on the state of the campaign. After my usual two or three hours sleep, I got up early to make phone calls and check the web site, took a shower, got dressed, and started outside toward open meadow where breakfast was being served. But the minute I stepped out the door, someone turned out the lights.

I couldn't see a thing. Everything was black. I slumped against the wall and steadied myself until slowly my vision started to come back, blurry and splotchy, but at least it wasn't total darkness anymore. This time, I didn't need a doctor to tell me what was happening. When you're battling diabetes, after your fingers and toes, your eyes are the next to go. For the first time, I was scared. What would I do if I lost my sight? I found a corner and waited until fuzzy shapes began to take form. By the end of the day, my vision had returned to normal. Afterward, I called Mike Ford and told him what had happened. But he was the only person I trusted enough to tell. Mike had been battling diabetes himself for years; it was one of the reasons he had quit day-to-day politics and one of the reasons he turned down the job of running the Dean campaign himself. I didn't want anyone else to know—especially Kathy. I knew she would make me quit. I couldn't quit. I was pitching the game of my life and there was no way I was coming out.

At least not yet.

A .357 MAGNUM

Where to start.

Describing how this miracle campaign got off the ground is one thing, but it might take five books to describe everything that conspired to rattle it to pieces just months later—the myriad cracks inside the campaign and the gale-force political winds we quickly encountered once we were off the ground. The truth is that we never really fixed the inherent problems in the organization that I saw that first day in Burlington, the same problems that caused Ford to say that I'd finally gone over the edge when I decided to take this job. Yes, we had built a dynamic, unconventional grassroots movement, but a presidential campaign was still waged, at least in part, in the conventional political world, and Dean for America had never mastered some of the most basic tenets of campaigning.

This was true at every level of the campaign, but nowhere was it more evident than at the top. This was the first presidential I'd ever worked on that didn't have a Campaign Chairman, a strong personality who could bring all the egos and personalities and ideas together. Howard simply didn't trust anyone unless they'd been around him for years, and the people who *had* been around him for years had no experience running a presidential. I don't blame Howard for this; one of his most remarkable traits is his strong sense of loyalty. But inevitably, leadership of the campaign remained splintered at the top between a half dozen people.

The deepest schism developed between Kate O'Connor and me. And because she was Howard's traveling aide, Kate was with him far more than I was and used her time with him to promote her view that I was leading Dean off in some dangerous, populist crusade. A *Time* magazine story after I'd left the campaign quoted an unnamed adviser (Hmmm, who could *that* be?) as saying that I was "turning a pragmatic politician" into "a fire-breathing liberal." As early as September, when we were cresting, she wanted me out. I had written a speech for the governor to deliver in Boston, but Kate and Howard decided that it went too far, so at first he agreed to only do three pages of it, and he only ended up reading a page-and-a-half, before switching awkwardly back to his stump speech. Afterward, a couple of staffers looking over Kate's shoulder saw her writing a message to someone else on her Blackberry—a personal communication device—that she wanted to get rid of me."[2]

I had been pleading with the governor for months to hire more experienced people, but once the campaign took off we just seemed to fall further and further behind. And as the fall turned to winter, we got into the awful rhythm of constantly fielding telephone calls from the press about some tiny controversy or another.

Dealing with the media in a political campaign is not for the weak of heart. It can be the most dangerous part of a campaign. One slip off message can be devastating.

---

[2] There was a hurried meeting about this at campaign headquarters, but Kathy, knowing that I was sick and already frustrated with the internal politics of the campaign, kept this incident from me until months later.

This is how it works in a presidential: The telephone is not a telephone. The telephone is a .357 magnum and every time you answer it and there's a reporter on the other end, what you're really doing is putting that gun to your head. Sure, most of the chambers are empty, but there's always at least one hollow-point in there, sometimes more. Every time you talk to a reporter, there is the potential that you'll blow your brains out. The higher your guy is in the polls, the more live rounds in the gun. The more unscripted your guy is, the more live rounds. Here we were on top of the world with the most improvisational candidate in history. Most days, there were more live rounds than blanks. That was why it was important to stay on message, so we weren't making things worse when we had to explain what the governor meant when he said that we wouldn't always have the strongest military in the world. Or so we weren't accidentally volunteering—as one person on the campaign inexplicably did—that yes, we were "worried" about our Internet supporters doing something stupid.

This is one reason Tricia Enright was such a great communications director, because she has a healthy respect for the telephone and knew how easily she could be talking to a reporter one second and wind up on the floor in a pool of her own blood the next. She's not timid, but she also knows when she's pushing her luck.

In our case, the gun only got heavier and heavier as the campaign wore on and Howard's once-charming insistence on speaking his mind ("You're going to find out that to my detriment and my credit I don't often think about the political consequences of what I say," he told Tim Russert on *Meet the Press*) became the very bullets that we had to duck every day.

"Your guy just said he has a hole in his resume in the area of defense and foreign policy. Does Howard Dean have a hole in his resume?"

How do you answer a question like that? Yes, he's got a hole in his resume? No, he doesn't have a hole in his resume. He must be mistaken. *Click.*

For months, I came out of this game relatively unscathed. And then, my own luck with the gun ran out, one day late in the campaign, when I was on CNN's *Crossfire* with Paul Begala and Tucker Carlson, and we were talking about a string of stunning endorsements, Bill Bradley and several members of Congress, that we had recently attracted to the campaign.

"Who's next?" asked one of the hosts. "Jimmy Carter?"

I could have smiled and shrugged. I could have said, *No, no*. I could have changed the subject. Instead, my answer was too cute by half. "You'll have to tune in Sunday and see."

This was at the very end, when you could hear the grinding of momentum turning heavily against us, and it didn't help when people tuned in the day before the Iowa Caucuses and saw Jimmy Carter cautiously escorting Howard Dean to the gate of Carter's house in Plains, Georgia, and the former president explicitly telling the press that he had *not* invited Howard Dean and that he was *not* endorsing Howard Dean, despite a rumor to the contrary—a rumor, of course, that I unwittingly had planted.

Boom! For months I had put the gun up to my head and came away smiling. Now here I was making my biggest fuck-up and taking that live round just days before the Iowa Caucuses.

But I'd had a bad feeling about Iowa long before that. On September 13, 2003, I went to Iowa for Senator Tom Harkin's Annual Steak Fry, one of those bellwether events in a presidential campaign. Right away, it didn't feel right. There was something in the air. Iowa was the worst state for us demographically; with a third of the caucus voters over sixty-five,[3] it wasn't exactly Internet City. And yet, a September Zogby Poll had us on top there, at 23 percent, to 17 percent for Gephardt and 11 percent for Kerry.

Still, it felt soft to me. As I talked to our organizers, I just felt that something was missing. Some measure of enthusiasm or confidence. In a campaign, your best supporters are called "ones"—those people who say they are definitely going to vote for your candidate, the people you can count on as sure votes. But when I asked our people on the ground in Iowa how many *ones* they had in their precincts, they did the single thing that pisses me off more than anything else.

I don't know, about five hundred.

"Don't ever do that," I said.

Do what?

In 1979, when I ran Jones County for Iowa, Tully, Ford, and Sasso made it abundantly clear to us Corn Stalkers that when they asked how

---

[3] Associated Press, *Gazette Online* (January 20, 2004).

many *ones* we had, we were never to give them a number ending in zero. A number ending in zero implied that we were estimating, which implied that we didn't really have a handle on our precincts, which implied that we hadn't worked hard enough, which implied that maybe we'd rather just be on a train steaming the hell out of Iowa. That day with Tully, Ford, and Sasso, I raised my hand cautiously. "Uh—what do I do if the number really does end in zero?"

"Lie," Tully said

Now, here I was, twenty-four years later, in a tight race in Iowa, and the softness of our numbers reflected a deeper problem. It wasn't that our people weren't working hard in Iowa. It was just a kind of mushiness, like a boxer who isn't in prime shape. A boxer who has lost his edge. He may look good now, but when the punches start flying, there could be trouble.

Paul Maslin was one of the few people on our side with presidential campaign experience, so I pulled him aside and suggested that we should think about getting out of Iowa. Leave the Caucuses for Gephardt and maybe Edwards to fight over. Pull out and concentrate everything on New Hampshire, where we were well ahead of the pack with 47 percent in an eight-candidate field.

Paul looked at me like I was just a little shy of going nuts—but he admitted the idea had some merit.

"I mean, what do we win here?" I asked. "So we show we can beat Gephardt? Big deal. We *know* we can beat Gephardt. Gephardt only plays in Iowa. The real race is with Kerry, and it begins in New Hampshire. Only one guy from the northeast is going to come out of New Hampshire. Why stay here, spend six million bucks and risk losing everything? Let's leave it to Gephardt and just get the fuck out. Sure, we'd take some hits for pulling out, but at least we'd be alive."

Paul gently reminded me that we were *ahead* in Iowa and that I needed to remember to sleep sometimes. One of my real regrets in the campaign is that, at that moment, Paul didn't talk me out of it. I talked myself out of quitting Iowa. No one else convinced me that we could still win Iowa that fall; I convinced myself that I was being paranoid. But the bad feeling wouldn't go away. In fact, it was about to get much worse.

THE ORATORICAL ABILITY OF A SIX-YEAR-OLD

No Democrat running for president had ever opted out of public campaign financing.

The reason for this is fairly obvious. The way our labyrinthine campaign finance laws have worked since 1976, a candidate can get matching funds if he raises a certain amount of money—but then he is restricted from spending *more* than that amount in any given state. So while matching funds can double a poorly funded candidate's base, taking the matching money means the candidate can't go over the cap. But Republicans have proven so adept at raising those $1,000 and $2,000 checks from the wealthiest Americans (who benefit most from high-end tax cuts and pro-business policies) they can raise two, three, four times the cap, while Democrats (relying on much smaller donations from its middle-class, working constituency) haven't been able to afford to forego the matching funds.

In the 2000 election, Al Gore raised more money than any Democrat in history, $49 million. But George W. Bush raised *$125 million*. Al Gore was the vice president of the United States, a brilliant, respected career public servant. And yet somehow a weak governor with five years of political experience and the oratorical ability of a six-year-old was able to raise two-and-a-half times as much money? How was that possible?

In 2000, almost half of Bush's money came from 59,279 donations of $1,000 (the maximum limit for donations in 2000), more than three times as many big donations as any candidate had ever received.[4] As Charles Lewis, executive director for the nonpartisan Center for Public Integrity wrote, "A contribution check of $1000 isn't something the average American can write; most often, those who open their checkbooks are lawyers, lobbyists, or the vested economic interests they represent who want something in return from the government." In effect, sixty thousand rich white guys determined who would be president for the rest of us three hundred million people. And this is not an aberration. Each year, the richest *one-quarter of 1 percent* of Americans make 80 percent of all individual political

---

[4] Charles Lewis and the Center for Public Integrity, *The Buying of the President 2004* (New York: HarperCollins, 2004).

donations. And corporations, the Republicans' primary source of funding, outspend labor, the Democrats' chief supporters, 10 to 1.[5]

Given the disparity in their funding, it's surprising Gore did as well as he did, winning the popular vote and only losing the election because of Florida's nostalgic return to Jim Crow-style electioneering[6] and the political intercession of five Supreme Court Justices.

In the fall of 2003, inside the Dean campaign, we realized for the first time that we actually had a shot at winning the nomination, and we began turning our attention to the Bush money machine—which had, if anything, become even more formidable than in 2000. The secret to Bush's fund raising is *bundling,* finding those corporate donors who can hit up their employees and country club friends for $1,000 and $2,000 checks. The Bush team calls people who could bring in $100,000 *Pioneers.* Donors who bring in $200,000 in bundled checks are called *Rangers.* According to an investigative project by the *Washington Post,* between 1998 and May of 2003, Bush raised $296.3 million and "at least a third of the total—many sources believe more than half—was raised by 631 people."

The king of these billionaire bundlers, until his company collapsed, leaving investors and employees high and dry, was Kenneth Lay, head of Enron ("Kenny Boy," as Bush called him) who brought in almost $600,000 for his old friend. As Lewis writes in *The Buying of the President 2004,* the key component of the "Pioneer System" is its nearly obsessive focus on making sure bundlers get "credit" for their fund-raising work.

> Now why would they want to receive credit? Clearly, many donors wanted first-in-line access to and influence with the prospective new administration. What is unusual about the Pioneer system is the unabashed directness of the transaction: You help us and we'll credit you and remember your loyalty and support later.

So what do the Kenny Boy Lays of the world get for their money? According to the *Washington Post*'s investigation, two of the five people

---

[5] Robert W. McCloskey, "Introduction," in *Profit Over People,* by Noam Chomsky, (New York: Seven Stories Press, 1998).

[6] According to Lewis et al., *The Buying of the President 2004,* 54 percent of the 180,000 ballots tossed out by Florida election officials were cast by black voters, even though they represented only 11 percent of the state's voters. And the *New York Times* reported that predominantly black precincts had more than three times as many ballots disqualified as white precincts.

named to the Federal Energy Regulatory Commission had been recommended by Lay. *In effect, for his $600,000 he got to choose 40 percent of the government agency that would regulate his industry and his corrupt company!*

> Of the 246 fund-raisers identified by the *Post* as Pioneers in the 2000 campaign, 104—or slightly more than 40 percent—ended up in a job or an appointment. A study by the *Washington Post,* partly using information compiled by Texans for Public Justice—found that 23 Pioneers were named as ambassadors and three were named to the Cabinet: Donald L. Evans at the Commerce Department, Elaine L. Chao at Labor and Tom Ridge at Homeland Security. At least 37 Pioneers were named to postelection transition teams, which helped place political appointees into key regulatory positions affecting industry.
>
> A more important reward than a job, perhaps, is access. For about one-fifth of the 2000 Pioneers, this is their business—they are lobbyists whose livelihoods depend on the perception that they can get things done in the government. More than half the Pioneers are heads of companies—chief executive officers, company founders or managing partners—whose bottom lines are directly affected by a variety of government regulatory and tax decisions.[7]

This fire sale on access and appointments goes all the way to the top. During Dick Cheney's five years as CEO of Halliburton, the huge Texas-based energy services company's employees more than doubled their donations to the Republican Party (about $1.4 million, compared with just $64,000 donated to Democrats), and the amount of money they were awarded in U.S. government contracts took a corresponding leap, from $1.2 billion to $2.3 billion. And it's taken another leap since Cheney became vice president. When he sat down with industry insiders to rewrite the country's energy and environmental policies, it's a sure bet that Halliburton officials were at the table. And they weren't the only ones. The oil and gas industries contributed about $17 million to Bush and the Republican National Committee and were rewarded with, among other things, the right to drill in national parks. In dozens of cases, lobbyists

---

[7] Thomas B. Edsall, Sarah Cohen, and James V. Grimaldi, "Building War Chest with Few Restraints," *Washington Post* (May 16, 2004).

who raised money for Bush were appointed to regulate the very industries they had once lobbied for. Meantime, the Republicans' control of
the House, Senate, and White House meant that corporations no longer
worried about spreading the money around. Pharmaceutical companies,
whose top five political donors gave more than $11 million to the Re-
publicans and less than $2 million to the Democrats, were rewarded with
a bill providing Medicare drug coverage for seniors, a bill that even its
proponents acknowledge is far more beneficial for drug companies than
for seniors.[8]

This was the rough political landscape we faced (and that John Kerry
now faces) as we looked at squaring off against Bush. Clearly, our only
hope was to raise enough money to compete, and that meant opting out of
public financing. We would be turning down $19 million in matching
funds, but it also meant we could go over the $45 million spending cap.
There was serious debate about this. I argued that we had tapped into the
political equivalent of a new Mother Lode with our Internet fund raising
and that our only chance was continuing to reach out to people who had
been disengaged from politics, that getting people to donate money was,
in effect, personally investing them in a candidate and in the system again.
Surveys had shown that if all eligible voters had cast ballots in 2000, Democrats would have won not only the White House, but both houses of
Congress, too. Yet low-income people are 47 percent less likely to vote
than high-income people.[9] The key was getting those low-income people
to believe that their vote mattered again. So even if the checks were for
ten or twenty bucks, in effect, raising money was our best hope of ener-
gizing the electorate. Our bat was the key to beating Bush, and it was
showing no signs of letting up. We'd initially hoped to raise $7.6 million
in the third quarter, but on September 20, with ten days to go, we had
$9.7 million and were close to breaking Bill Clinton's one-quarter record
of $10.3 million. So we set—and topped—a goal of $5 million for the last
ten days, for a total of $14.8 million for the quarter. Unfortunately, some
idiot whose math skills apparently peaked in junior high school had looked
at the $9.7 million and rounded up, setting the public goal at $15 million.
So that even while the media marveled at our fund raising, we had to

---

[8] Lewis et al., *The Buying of the President 2004.*
[9] Thomas E. Patterson, *The Vanishing Voter* (New York: Alfred A. Knopf, 2002).

answer inane questions about whether we were worried that we'd come up $200,000 short of our goal. Oh, by the way, that idiot was me.

Our fund-raising was proving to be one of the most amazing parts of this story. After his record-setting third quarter in 1996, Clinton raised only $6 million the following quarter. But our fourth quarter started out even better than our third had (in the end, we would top our own record with $15.8 million). It was clear that at the rate we were raising money and adding new supporters, our best chance at winning everything was to forego the matching funds in order to be able to exceed the spending caps.

Still, some people inside the campaign worried that opting out of public funding sent the wrong message, betraying our populist roots and our candidate's commitment to campaign finance reform. ("Hell, we *are* campaign finance reform," I used to say.) So we decided this would be another good test of our open-source campaign. We'd put the question out to our supporters. We'd hold an online referendum.

This was a critical moment, the first time in history that a campaign had put its future in the hands of its own supporters. If you put something like this to a vote, you have to abide by the results. If they told us *not* to opt out, we would have had no choice but to do as they said and—had Dean been the nominee—we could have faced a general election in which we were outspent two- or three-to-one.

I was on the road when we posted the referendum, so I called Nicco the first day to ask what the results looked like. He told me that about 100,000 people had voted and more than 80,000 wanted us to forego the federal funds, to take the chance that we could raise the money with small donations from regular Americans. We had decided that if people voted to opt out, we'd ask them for a donation, so I asked Nicco how much money we'd raised that way.

"Two hundred forty eight thousand bucks," Nicco said. (Actually, having heard my rant about how I never wanted to hear a number end in zero, I'm sure Nicco probably said something like: two hundred forty-eight thousand *and six*.)

"No, no, no, no," I said. "You can't have a hundred thousand people vote on this and only raise two hundred forty eight thousand dollars. It makes no sense." I told him to check again.

Nicco assured me that the numbers were correct.

I landed in Manchester around midnight, drove to Burlington, and got to campaign headquarters at about 3:00 in the morning. I immediately saw the problem: When people voted to opt out, they were directed to a new web page, where they were thanked for voting and told that by rejecting matching funds, we were kissing eighteen million dollars goodbye, and that they could help us make up for that eighteen million by contributing. Fine. The problem was that the web team had put the donation request at the *bottom* of the page. So people read the *Thank You* part of the page and missed the donation part, which they only saw if they scrolled down.

One of the most important rules of communicating on the Internet is this: *Put it on one page.* And if it's important, put it on top of the page. Most people will click away from the page rather than scroll down.

This was one of those days in which the web team hadn't slept in seven or eight days (who am I kidding, every day was like that) and they had stared at their computer screens for so long, they completely missed the problem. We quickly rearranged the Thank You page to put the solicitation up high, and the next day another 100,000 people voted, but this time *$4.5 million* came in. We had missed out on four million bucks the day before.

But even with the occasional $4 million mistake, by all accounts, we were flying high.

The first week of October, I called a meeting with my partners, Steve McMahon and Mark Squier, the pollster Maslin, the deputy campaign manager Tom McMahon, and my wife, Kathy. We were ahead in fund-raising, ahead in national polls, ahead in Iowa and New Hampshire. The media was practically handing us the nomination; a headline in Roll Call, the insider publication of Capitol Hill said, "It Ain't Over 'Til It's Over, But It's Probably Over." Stories began speculating on Dean's running mate. And personally, I was getting a lot of the credit for this. Reporters made me out to be a high-tech version of Karl Rove, George Bush's deviously brilliant chief strategist. Bob Shrum, my old boss, who is not one to lavish praise, grabbed me at the Harkin Steak Fry and told me that I was running "the most brilliant campaign" he'd ever seen. Twenty-four years after my first presidential campaign, I really *was* having the game of my life, taking the ultimate dark horse candidate from out of nowhere to the brink of the Democratic Party nomination.

I suppose that's why my partners and Maslin and Kathy looked a little shocked at that meeting in early October when I announced that I was going to quit.

THE CORONATION

Iowa just didn't feel right. And to fix Iowa, we were diverting much of the energy that had been driving our Internet campaign. We were in danger of flattening out. Meanwhile, the Dean road show was a total disaster—a new gaffe every day. The other Democrats were starting to pummel us, as were the conservative special interests, and we were still making fresh-man mistakes. Ten months after I'd first complained about it, we still didn't have enough experienced people on board. While we had the most energetic group of staffers and volunteers I've ever seen on a political campaign, we were nearing treacherous water and we badly needed experienced people. And the worst part was that I didn't think I could do a goddamn thing about *any of this.* Without the checkbook, without Dean's trust, I had no real authority. And with my health declining, it just wasn't worth it—it wasn't worth killing myself over something I had no chance to fix.

"This thing is going to crash in an Iowa cornfield and I can't do anything to stop it," I said. "I've been trying since March and it just ain't gonna happen. I should leave and let them get someone else in here. Maybe they'll let someone else fix it."

They all tried to talk me into staying. They said the campaign needed me, that a lot of people had gotten involved because of me. "After all our work, we're on top now," Steve told me. "It'll break your heart to leave now, Joe."

"It *will* break my heart," I said. "But it'll break it even more to stay and watch it go down and know that I couldn't do anything to stop it."

I began a countdown. "Ten more days and I'm gone. Eight more days and I'm gone. Six more days . . . ."

My deadline came and went. I started a new one. "Ten more days and I'm gone." We were slowly bringing in a few more aces, like Gina Glantz and Jay Carson, who had worked on Bill Bradley's presidential campaign; Andi Pringle, who had been Carol Mosely Braun's campaign manager; and a little later, Jon Haber, the gregarious old hand I'd worked with on the

Gephardt, Mondale, and Ted Kennedy campaigns. Even Mike Ford came back. It looked like things might get better.

But they were about to get a lot worse, culminating on December 9, a day that looked like the high point of the campaign, but turned out to be the beginning of the end.

It started a couple of days earlier, with a screaming phone call from our political director, Paul Blank, the latest person to go to Iowa. "I'm out here trying to fix this thing, to get everyone on the same page and now all of a sudden I'm supposed to build a twelve-hundred-person crowd in Cedar Rapids with no clue why I'm doing it?"

I told Paul I had no idea what the hell he was talking about.

He said the governor had called him and told him there was a big surprise to just make sure a lot of people were there.

That led me to Todd Dennett, my new scheduler. I liked Todd, but I had been fairly surprised to find out that I had a new scheduler since the governor had simply hired him out of the blue one day.

I asked Todd if he knew anything about a campaign stop in Cedar Rapids, and he just gave me a quizzical look. "Does this have anything to do with the governor asking me to find a sixteen-seat Gulfstream for New York that same day and he told me not to tell anyone about it?"

"Maybe," I said. This was one of those moments. Talking to the scheduler I liked but didn't hire about a mystery crowd being built in Cedar Rapids and a jet being chartered in New York—I could've sworn that when I'd gotten up that morning I was the *manager* of this campaign, and yet I had absolutely no clue what was going on. (Meantime, my partners and wife couldn't figure out why I wanted to quit, and some time in the near future people would start wondering if a guy once considered one of the premier organizers in the Democratic Party—me—had gotten a frontal lobotomy, because suddenly it looked like he couldn't manage himself out of a paper bag.)

I called Howard. "You were my very next call," he said. He seemed as mad at Paul for telling me as he was embarrassed at having been caught leaving his ostensible campaign manager out of the loop.

"What's going on?" I asked.

"I can't tell you," he said.

"Governor—"

"We're getting a big endorsement."

"Who?" I asked.

"I can't tell you. You'll just have to trust me."

Trust. The irony was not lost on me. I had been busting my ass to get much-needed labor endorsements for Dean and we had just scored an unlikely coup: endorsements from the two biggest unions in the AFL-CIO, the 1.4-million-member American Federation of State, County, and Municipal Employees (AFSCME) and the 1.6-million-member Service Employees International Union (SEIU), whose members included doctors, nurses, health aides, and hospital janitors. Andy Stern, the president of SEIU, and Gerry McEntee, the powerful, legendary president of the AFSCME, had shown real leadership and courage in moving to support Howard, leadership that in my view was the thunderclap that really woke up the Democratic Party. With the unions signed up, soon members of Congress began to come around.

And now, Dean had some big endorsement and he was keeping me in the dark about it.

Howard explained that the endorsement would take place in Harlem and then we'd fly to Cedar Rapids. Harlem. My first thought was that Bill Clinton's office was in Harlem. But Clinton's centrist wing of the Democratic Party seemed dead set against a Dean candidacy. There was no way it could be Clinton, could it?

"I can't tell you," Dean said.

Tricia and I had to call the campaign reporters and tell them to come to a secret announcement in *Harlem,* and when they put two and two together and asked if Clinton was endorsing us, we had to say—Well, no—at least, we didn't think so.

"What do you mean, you don't think so. You mean, his own campaign manager doesn't know?"

No.

I found out the same day as everyone else that it was Al Gore.

Howard had been keeping in touch with the former vice president and had asked him to read a major foreign policy speech that he was about to give. Gore, who was in Japan, called him about the speech and toward the end of the conversation matter-of-factly told Howard that he would endorse him. They purposefully designed it as a surprise, a bombshell to be delivered at a standard Dean campaign event—an event that just happened to be in Harlem on December 9.

To their credit, it was a great day; it really did feel like a coronation of some kind. Still, it turned out to be the beginning of the end for us, the moment our insurgent campaign ended and our short, uncomfortable stint as overwhelming frontrunner began—with a target squarely between Howard's shoulder blades.

Just days after Gore's endorsement, Saddam Hussein was captured in Iraq. The next day, Dean was in Los Angeles to give that foreign-policy speech. During the speech, he offered a single line that he'd added to the text on the drive over: "The capture of Saddam Hussein has not made America safer."

Of course, it turned out to be true, even prophetic. But in the world of broadcast politics, truth and two dollars will buy you a cup of Starbucks coffee.

CHUM IN THE WATER

PHONY INTERVIEWER: "What do you think of Howard Dean's plan to raise taxes on families by $1,900 a year?"

PHONY IOWA MAN ON STREET: "What do I think? Well I think Howard Dean should take his tax-hiking, government-expanding, latte-drinking, sushi-eating, Volvo-driving, *New York Times*-reading, body-piercing, Hollywood-loving, left-wing freak show back to Vermont where it belongs. Got it?"

As campaign spots go, it wasn't the subtlest in the world. And it wasn't even close to the worst we would face that winter.

For months, ever since our summer charge, we'd known the other campaigns were going to hit us. Hell, they *told* us they were going to hit us. The August 11, 2003, issue of *Time* had quoted a "strategist for another Democratic contender" as saying that they were about to come after Dean, but that, "it's kind of like the Mafia. Everyone wants another family to hit him. You don't want to bring blood into your own house."

And it wasn't just the other campaigns. After twenty years of watering down the Democrats' message to play to the center, the party leadership was scared to death of a Dean candidacy. Senator Evan Bayh, chairman of the Democratic Leadership Council released a statement that read: "It is our belief that the Democratic Party has an important choice to make. Do we want to vent or do we want to govern?—The Administration is being run by the far right. The Democratic Party is in danger of being taken over by the far left."[10] Every step of the way, we had to fight our own party leadership. Earlier, the debates had been set up in formats that hurt "lesser candidates" like Dean. Kathy had been doing debate liaison with the DNC, and I told her that we needed to make sure we gave it right back to them. "If they tell you it's red, you tell them to fuck off, we want green. If they want a podium, we don't want a podium, if they say it's up, you tell them we want down." When Josh Wachs, the executive director from the DNC, called, people used to gather in my office to hear me berate them. I later found out that people at the DNC used to gather at his end to hear the action, too. In the end, I came to respect Josh a great deal. He had a thankless job, herding cats, and I suppose I didn't make his job any easier.

That winter, our favorable ratings in Iowa took their first dip, as the other candidates hammered Dean for his Saddam comment and for everything else under the sun. We were caught in a crossfire between our own party and Republican political action groups like Club for Growth, which had put up the left-wing freak show spot. But all of this was a picnic compared to what happened after Gore's endorsement. The other campaigns had been circling like sharks, occasionally coming in to take a bite. But the Gore endorsement was like chum in the water.

Every single campaign came after us. They all dumped their oppo research files on us, unloading everything they'd been saving up. Even Al Sharpton, who didn't have oppo research, came after the governor—in Sharpton's case, for not having any black people in his cabinet in Vermont. (The *Village Voice* did a scathing investigative piece that strongly suggested Republicans were feeding Sharpton information and providing some funding.) It was as if the other candidates all got together and said, we have to kill him and we have to kill him now. What followed was a cascade of *push polls* (phony polls in which negative innuendo and slander are disguised as survey questions), savage attack ads (nothing on the screen but a picture of Osama bin Laden while the announcer talked about Dean), and telephone harassment of our *ones,* the supposedly committed Dean voters. Mysterious political action committees like Americans for Jobs, Health Care and Progressive Values appeared at the last minute,

---

[10] Mark Singer, "Running on Instinct," *New Yorker* (January 12, 2004).

attacking us with ads that had Dean in bed with the NRA, Newt Gingrich, and George Bush.[11]

This has happened to outsiders in politics before, of course. After he surprisingly won the New Hampshire primary and threatened to derail George W. Bush's money train, John McCain was similarly thrashed in South Carolina with coded attack ads and push polls, phone calls and flyers alleging that McCain had "a black baby," was being helped by "a Fag army," and had been brainwashed in Vietnam to destroy the Republican Party. One of McCain's staffers accurately called it "a jailhouse rape."[12] Even Jimmy Carter felt both barrels of his own party turn against him when he shocked the world in 1976 by winning Iowa. But this was the first time a candidate had been eviscerated *before a single vote was cast*.

Dean was knocked for flip-flopping on everything from Medicare to Social Security, for being too liberal *and* for being too conservative. ("These guys can't make up their minds whether I'm Newt Gingrich or George McGovern!" he said at one debate.) And under this full-frontal assault, he sometimes made it worse with his insistence on saying whatever popped into his head. Like when he volunteered that a conspiracy scenario that Bush had known beforehand about the September 11 terrorist attacks was "an interesting theory." It only got worse when he tried to get out of it by telling reporters, "I did not believe the theory I was putting out."[13]

And then there was the decision to seal his records that led to the meeting in Burlington in which he admitted that he'd never expected to be in a position to win. Even candidates who weren't campaigning in Iowa took shots at us on that one. "Everybody ought to be open about what they've done in public life," said Wesley Clark the day he released all of *his* public records. Every day it was the same: Here's the attack, what's your response? Here's the attack, what's your response?

The worst thing that can happen to a candidate is when he stops defining his own candidacy. After Gore endorsed Howard Dean, alarm bells went off in every newsroom in the country and in every other campaign headquarters. The campaign alarms said, *we've got to kill this guy now, or he's going to be the nominee.* The press corps alarm said, *this guy*

---

[11] This group conveniently filed its incorporation papers too late for disclosure before the Iowa Caucuses, but turned out to be made up of Kerry and Gephardt supporters.
[12] Lewis et al., *The Buying of the President 2004.*
[13] *Washington Post.*

is about to be the nominee. We've got to put him through that ringer that every future nominee goes through. That's our job. For a month, those two forces beat against us.

All along, the media had blithely referred to some odd strand of conventional wisdom that Howard Dean was too intense to be president. In two feature stories over the summer, the *Washington Post* had described him as: abrasive, flinty, cranky, arrogant, disrespectful, yelling, hollering, fiery, red-faced, hotheaded, testy, short-fused, angry, worked up, and fired up.[14] Those last weeks before the Iowa Caucuses, the media was on the lookout for that *one story:* the testy insurgent not yet ready to lead, cracking under the pressure. And so, when a week before the caucuses, Howard told a persistent heckler to "Sit down. You had your turn—it's my turn now," it fed into the story the media had wanted to tell all along, never mind the fact that Dean hadn't lost his temper even once in the campaign, and that the guy who had been haranguing the governor for ten minutes was a Republican.

Al Gore's endorsement was a great moment for the campaign. I admire his courage and resolve for embracing Howard Dean. But because it came more than a month before the Iowa Caucuses, it gave the rest of the field time to come after us with everything they had. In a June 2004 *Atlantic Monthly* article, Joshua Green wrote about the onslaught:

> By the time Gore endorsed him, on December 9, Dean's victory in the upcoming primaries seemed assured.
>
> That same week Ben Holzer, the research director for General Wesley Clark's campaign, arrived with (Chris) Lehane, who was then working for Clark, in Washington, D.C., for a series of visits to the major television networks, newspapers, and newsmagazines. They toted a three-ring binder that contributed as much as anything else to Dean's rapid demise. The Clark campaign had classified the stories in it as singles, doubles, triples, or home runs, based on the damage they were expected to inflict. Holzer and Lehane offered producers and reporters exclusives on many of these stories with the proviso that if they were not used quickly, they would be handed to a rival. In the hypercompetitive world of political journalism this pretty much guaranteed swift airing or publication.

---

[14] Mat Gross, "Dean for America Blog" (January 2004).

It wasn't just Clark, of course. All of the campaigns came after us this way. And under that blistering attack, the cracks in our campaign organization became gaping holes.

I would love to be able to say that if I'd known about the Gore endorsement, I would have told Dean to wait until a week or ten days before the caucuses, to buy us some more time as the outsider, give us time to fix some of our internal problems, and to make better use of the momentum going into Iowa. But honestly, I don't know if I'm that smart. What I do know is that I never got the chance.

Of course, in the end, it doesn't matter. We could've gotten the Gore endorsement later, still lost in Iowa, and be complaining that it came too late. You can't know. A campaign is made up of thousands of tiny decisions and actions, all but a few of which you can second-guess. We had no more or no fewer internal problems in the Dean campaign than any other organization trying to squeeze one guy to the top of the world. Did Howard make mistakes? You bet he did. This was his first real campaign. Did his staff make mistakes? Every day. Did I make mistakes?

More than I will ever know.

But God help us if a mistake by Howard Dean or Joe Trippi masked the real victory of the 2004 campaign: the way people heard the words "You gotta believe" and they did—in themselves, not in some knight in shining armor promising to save them.

In the end, the infighting, the attack ads, the gaffes, none of it explains why John Kerry became the democratic nominee.

Eight candidates go in; only one can come out. We came up short, along with six other candidates. The difference is that no one is looking at Dick Gephardt or John Edwards and saying that the fact that they lost proves that the old style of campaigning doesn't work. Whatever mistakes we made, whatever didn't go right, whatever old-school politics contributed to our fall—none of that takes away from all that we were able to accomplish. The Dean campaign did not lose because we relied too much on its populist Internet supporters. The truth is this: The only reason the Dean campaign even got close enough that it mattered was because of those people.

10

THE END
Murder/Suicide, Harley Davidson, and Going Home

THE MINUTE I saw the spot I knew.

We were dead.

The Dean for America campaign had been hammered for months from all sides, and when our numbers began to flatten, we'd had no choice but to hit back. In October, a Zogby poll had Dean and Gephardt in a dead heat in Iowa, at 21 and 22 percent, with Kerry well back at 7 percent. Two months later, the *Los Angeles Times* ran a poll that showed Dean ahead in Iowa with 30 percent, Gephardt with a slipping 23, and Kerry rising at 17. But, as the poll pointed out, Iowans were very "fluid" in their choices; their minds weren't entirely made up (this was the softness I'd felt three months earlier). And as the day of the caucuses approached, the bad news just kept coming: An old video surfaced showing Dean on Canadian TV criticizing the Iowa Caucus system—and once that tape hit, we really started to bleed. One week before the vote, our internal numbers put Dean's favorable rating at an all-time low in Iowa and had the race as a dead heat between Dean, Gephardt, and Kerry, with Edwards only a few points back.

We saw no choice but to be aggressive and so we put up a television ad focusing on the issue that had served us best in Iowa—the war in Iraq.

Until that point, the TV spots put up by all the campaigns had generally followed the logic of a multicandidate race—if you're going to hit one, hit them all. The rationale goes like this: In a four-man race like we had in Iowa, if you hit just one guy, the votes splatter to the other two. The votes never come back to you after a negative ad. So if we nail only Gephardt, he loses voters because of the ad, we lose voters for going negative, and those voters have nowhere to go but to Kerry and Edwards. If you make the mistake of hitting just *one candidate* in a four-person race then you allow the other two to run out in the clear.

That's why we included all three guys in our spots, which we also designed to be more focused on defining Howard than on slamming the other guys:

ANNOUNCER: Where did the Washington Democrats stand on the war? Dick Gephardt wrote the resolution to authorize war. John Kerry and John Edwards both voted for the war. Then Dick Gephardt voted to spend another $87 billion on Iraq. Howard Dean has a different view.

DEAN: I opposed the war in Iraq and I'm against spending another $87 billion there.

Clearly, the ad was toughest on Gephardt (he would end up losing fifteen points in fifteen days) but it followed the logic of a multicandidate campaign by going after the other two guys as well. And that's what Gephardt had done with most of his earlier spots, too, primarily blasting us on the issue of trade, but making sure he spread the misery to John Kerry, Wes Clark, and Joe Lieberman at the same time.

But this time, the Gephardt campaign did something inexplicable—and to this day, I still can't decide if it was incredibly cynical or amazingly stupid:

ANNOUNCER: "Did you know Howard Dean called Medicare 'one of the worst federal programs ever'? Did you know he supported the Republican plan to cut Medicare by $270 billion? And did you know Howard Dean supported cutting Social Security retirement benefits to balance the budget?"

With a week to go, and Gephardt bleeding from our Iraq ad, his campaign made the decision (or the mistake) to take Dean down with him. Ignoring the rule of a multicandidate race, they put up these ads aimed *only at us.* And as it turned out, these ads were even more deadly than they'd imagined: They took us *both* out of the race. In the wake of those spots—and the media coverage that we were in a negative ad war with Gephardt—both campaigns cratered, hitting their lowest poll numbers with just days left to go. We got Gephardt's people to call a truce and both sides pulled their ads, but by then the damage was done. Gephardt had committed a murder-suicide, with us as the murder victim. Tens of thousands of votes were splattered off Dean and Gephardt and those votes had nowhere to go but to Kerry and Edwards—who were surging anyway, and who got last-minute boosts by looking as if they were above the fray (Edwards even took credit for our truce). Any chance we had to turn it around in Iowa was gone. Combined with the mistakes we'd been making, this meant we weren't getting off the mat. By Friday, a week after the *Los Angeles Times* poll and three days before the caucuses, our internal polling showed us trailing Kerry, 36 percent to 18 percent.

A month earlier, it had been Kerry on the ropes. The Massachusetts senator had figured to skate over Iowa, likely giving it to us or to Gephardt and concentrating his energy on New Hampshire. But under the leadership of Karen Hicks, a savvy native and political veteran of the Granite State, we'd taken such a commanding lead in New Hampshire that Kerry had been forced to come back to Iowa in a last-ditch effort to stay in the race.

Now, John Kerry was looking like the guy to beat.

But we weren't giving up. The Dean campaign would make one last stand—3,500 committed volunteers, mostly people driven by our incredible Internet support all over the country, came to Iowa to knock on 200,000 doors and spread Dean's message of empowerment. They wore orange stocking caps and called themselves "The Perfect Storm" (after the posting I'd made in May). It would be one last show of populist force by the most dedicated group of supporters I'd ever seen. One hundred eighty came from Texas. One hundred fifty from Pennsylvania. Twenty volunteers coming from California ran into a kid from Pennsylvania and talked him into stopping in Iowa to help them knock on doors for Dean. Three American expats came from Japan to help reclaim their country. It was humbling, how hard these volunteers were willing to work. I went out with them, into the cold, knocking on doors and trying to undo the damage from our month of freefall, capped by that final Gephardt assault.

But the other campaigns went to great lengths to make the Dean volunteers in their orange hats seem like mettlesome outsiders and our reception was almost as cold as the weather.

It was simply too late to do anything about Gephardt's last spot. It wasn't lost on me that after eleven months of showing how the Internet would be the technology of future elections, in a few days we'd been taken down by the technology of past elections—the TV attack ad. I had no idea that worse was still to come.

By the morning of the Iowa Caucuses, Monday, January 19, 2004, we knew that it looked bad. We were just hoping to hold on and finish second.

We didn't. Kerry and Edwards blew past us and we just never moved off of our bottomed-out 18 percent, a devastating third-place finish. Gephardt, who had led Iowa for almost a year, shot himself down to 11 percent.

Of course, everyone *thinks* they know what happened next. That old loose cannon Howard Dean, finally unhinged by his poor performance in the caucuses, stepped on-stage in a Des Moines ballroom and went crazy, screaming out a list of states almost at random.

Only that's not what happened. First, let me say that Howard Dean was not a loose cannon. He was not *flinty* or *prickly* or any of the other code words the media used to insinuate that he lacked the temperament to be president. For more than a year, under the intense pressure of a long-shot presidential campaign, Howard Dean was a serious, good-natured, self-possessed leader. And if anyone could have expected to see that side of Howard, it would have been me, his dispositional opposite. But Howard yelled at me exactly one time, and it was for a good reason, when I told field director Tamara Pogue that she didn't need to constantly give him updated numbers on his support. Howard did not go around yelling, or snapping at people, nor was he always on the verge of erupting. But this was a story the media was determined to report, even though it simply wasn't true.

Think about that clip of his concession speech and how many times you saw it repeated over those forty-eight hours after the Iowa Caucuses. Now ask yourself this: Did you ever see the audience? Did you ever see the people Howard Dean was speaking to? No. You didn't see a standing-room only crowd—more raucous than most *victory* parties I've been to—exhausted-but-exhilarated people who had come from all over the country to be involved in this thing, true believers who were exhorting their leader to go on believing right alongside them. You didn't see this because they didn't show it to you. You didn't see them screaming and hooting and leading Howard on because they only showed a few seconds of the clip. You didn't see the way these people lifted him up, the way they lifted us all up with their energy and their refusal to take this *loss* as a *defeat*. If there was any question about whether the Dean campaign would continue to fight, the people in that room answered it:

You know something? If you had told us one year ago that we were going to come in third in Iowa, we would have given anything for that. . . .

And you know something? Not only are we going to New Hampshire, we're going to South Carolina and Oklahoma and Arizona and North Dakota and New Mexico! We're going to California and Texas and New York! And we're going to South Dakota and Oregon and Washington and Michigan! And then we're going to Washington, D.C. To take back the White House!

We will not give up. We will not give up in New Hampshire. We will not give up in South Carolina. We will not give up in Arizona or New Mexico, Oklahoma, North Dakota, Delaware, Pennsylvania, Ohio, Michigan! We will not quit now or ever! We want our country back for ordinary Americans. And we're going to win in Massachusetts! And North Carolina! And Missouri! And Arkansas! And Connecticut! And New York! And Ohio!

Most of all, let me thank you from all over America, coming here to change this country. We haven't seen this in thirty years.

This is the changing of the generations, the passing of the torch to the new generation. It is your generation. And it's your generation that's fueling this campaign because you know that the half trillion-dollar deficit this president is piling up is going to be billed to you and your children, because of the terrible damage this president is doing to the environment are going to be things that you're going to have to live with.

And we're going to change that. And you have the power to change that. And we are starting right here tonight.

I remember, at the time, thinking that Howard had given a great boost for our supporters in the room, yet I knew this was going to play badly on TV. (What I didn't know was that it would play and play and play and *play* badly . . .) Howard wasn't thinking about the TV cameras.

It was all about those people. It was *their* energy he was channeling. It was *their* campaign. That's what the broadcast media never got—the cameras stayed on the candidate that night in Iowa, and because of that, they totally missed the campaign.

Every time I see John Kerry on television, I am struck by the same thought. Whether he is snowboarding or playing hockey or wearing his leathers, sitting on a Harley Davidson motorcycle, the message seems to be the same: *Aren't I amazing?* His campaign spots, with their focus on his laudable service in Vietnam, tell the same story: *Aren't I amazing?* It's no different with the other Democratic candidates, or with George W. Bush, the "compassionate" conservative. (Isn't this something like a vegetarian shark?) All of them had their version of Kerry's Harley. *Look at me. Aren't I amazing?* In a profile the *New Yorker,* Mark Singer talked about Dean's rejection of this sort of politics:

> [Dean's] principal rivals—John Kerry, Wesley Clark, John Edwards, Richard Gephardt, Joseph Lieberman—all subscribed to the focus group-tested orthodoxy that an ample measure of ostensibly intimate disclosure was the sine qua non of effective communication with the electorate. So Kerry alluded incessantly to his exploits in Vietnam; Clark spoke of the pride he felt in his military career; Edwards and Gephardt emoted about having been shaped by their fathers' humble livelihoods (textile-mill worker, milk-truck driver)—but Dean was having none of it.
>
> Dean's abstinence from personal revelation was just one way that he had discombobulated the Party leadership's assumptions about how to reclaim the White House.

This is the difference between the Dean for America campaign and every other presidential campaign of the past twenty years. Every other candidate has started by saying—*Look at me. Aren't I amazing?*

But every time Howard Dean got up to speak, every time his campaign staff got on the web to blog, the message was *Look at you. Aren't you amazing?* And they were: 600,000 amazing people committed to a new democracy.

One of the high points of every election cycle in Iowa is the JJ Dinner—named for the party fathers, Jefferson and Jackson—at Veterans Memorial Auditorium in Des Moines. It is one of those Iowa traditions nearly impossible to explain, a blend of political pep rally, fund-raiser, straw poll, and candidate beauty pageant in front of thousands of screaming, chanting, discerning Iowa Democrats. At the 2003 event, the other candidates did what candidates always do: They preened and posed on the floor, shaking hands with the big wigs and sitting at the $1,000 tables with the big donors. But when Howard was introduced, no one could find him at first. It turned out he was in the balcony, in what one person called "a sea of Dean," with all those thousands of people who had paid forty bucks to be a part of this. Our supporters went ballistic. We didn't call him "People powered Howard" just because it rhymed. This was our campaign.

You see, the revolutionary idea behind Howard Dean's campaign was not the use of the Internet. It was not his lonely opposition to the war in Iraq. The revolutionary idea of this campaign was that 600,000 people could rise up in the blink of an eye, and in a matter of months they could change the political landscape. In a matter of months they could do more than power a long-shot presidential campaign. They could begin the difficult task of restoring fairness to our political system. They could matter again.

And while a lot of mainstream politicians who feared a Howard Dean campaign could sleep peacefully after what happened to Howard in Iowa, what they forget is that those 600,000 energized Americans are still out there. In the next election, they will number two million. Or four. Or ten. And now that they've had a taste of the power that's been held back from them all of those years, they are going to want more.

THE BACKBONE TRANSPLANT

We flew to New Hampshire right after the Iowa count, landed in the early morning hours for a huge airport rally, then zipped off to our hotel, and fell asleep for a couple of hours. None of the reporters traveling with us said anything about Howard's remarks, other than noting that the crowds had been wild. We had no idea there was any controversy until we woke the next morning to a surreal hailstorm of criticism and scorn over, of all things, a concession speech. You almost had to remind yourself: this guy wasn't arrested for assault. He didn't steal money. He wasn't caught in a sex scandal. He was—too enthusiastic? All of the cable news channels repeated the "scream" every hour on the hour for two straight days. Ostensibly objective news anchors and reporters called it "bizarre," "scary," and "rabid." If an attack ad on Medicare could derail an entire campaign in Iowa, imagine what this wall-to-wall mockery could do.

After killing us this way, it was laughable to read in newspapers and see on television the severe pronouncements about the way we lost Iowa. Caught off guard by Howard Dean's rise, the traditional media seemed to take great pleasure in his campaign's freefall (which, in turn, helped speed its descent). The conventional wisdom among many reporters was that Dean for America was an Internet bubble bursting before the campaign. That it could translate its online success into real-world votes was just another over-hyped technology start-up.

This is ludicrous. A large part of our base may have come to us on the Internet, but you don't get to 32 percent in the Iowa polls only by using the Internet. You don't get to 47 percent in New Hampshire without having more than bloggers. You don't get 600,000 supporters at a time when other campaigns have a fraction of that number only by using the Internet. You certainly don't raise $50 million by only putting up a web site. We *had* translated Internet momentum into a front-running campaign. You can't attribute the campaign's fall to the limitations of the Internet. We had left those limitations far behind. No, what took us down were those two old standbys—our own mistakes and, most of all, the hungry, fickle beast of broadcast politics.

But here is the biggest misconception about the end of the Dean for America campaign: the idea that it ended.

When I began, we had no illusions about how hard our task would be. Remember, in thirty-four years, only *two of sixty-three* Democratic candidates became president, and neither of them started as far back as Howard Dean started. We knew that going in. In that room above the Vermont Pub and Brewery back in January, we made one pledge to ourselves and to those 432 supporters. Yes, we were probably going home early, but we would leave our mark. We weren't going to be one of those forgotten campaigns. Whatever happened, years from now people would remember Dean for America.

For the next year, we ran point for a pack of frightened candidates, saying all the things they were afraid to say, framing the debate, practically writing the party's message. We taught the Democrats how to be the opposition party. We showed them that Bush was vulnerable. We showed them that you didn't have to play by the old rules. And we showed them how to use the only tool capable of returning democracy to our system: the Internet. By summer, the Democratic presidential campaign was being waged on our terms, using our language. And our impact wasn't just in the presidential race. In part due to Howard Dean, members of Congress and the media and regular Americans began to realize that they could criticize the war and not be unpatriotic—in fact, the most unpatriotic thing they could do was remain silent while President Bush lied about his motives for going to war. The truly American thing was to object to the way our political system and our ideals were being sold out from under us. As *Washington Post* columnist E. J. Dionne wrote two days after the Iowa caucuses:

> What Democrats needed after their disastrous losses in the 2002 election was a backbone transplant. The party's rank and file were clamoring for less timidity in confronting George W. Bush. The yearning was not just—or even primarily—about the war in Iraq. For most, it simply meant having leaders who stopped looking over their shoulders and checking Bush's popularity ratings. Democrats were sick of intimidation and capitulation.
>
> The good doctor Dean answered the need and he soared. What he did not count on is that Democratic presidential candidates are a teachable species. They made adjustments.

They certainly did. In that brief, wonderful period when our campaign soared, the other candidates—most of whom had voted to go to war—decided they were fervently against it, too. Suddenly, Edwards' and Kerry's campaigns (funded predominatly by $2,000 donations) aired TV ads that sounded suspiciously like Howard Dean's rejection of moneyed special interests. By the eve of the Iowa Caucuses, the other campaigns had stolen everything we did and everything we said, from our message to our use of the Internet—everything. When John Kerry's fund-raising "hammer" went up on his web site, all we could do was shrug. This is what happens in a campaign. There are no patents in politics. But for the Dean nation, the end result of this grand theft was that even when our candidate began to fade, our message continued to resonate through the campaign. If John Kerry ends up in the White House, it will be in large part because of Howard Dean. Howard's message, his willingness to take on the White House, his use of the Internet—these are the things powering the

Kerry campaign. And while no pundits will talk about this because it doesn't fit their pretty picture of a political horse race, the sad truth is that if the Dean campaign hadn't opted out of public financing, giving Kerry the political cover that he needed, there would be no race for president in 2004. Bush would have a four-to-one spending advantage and the Democrats would be talking about 2008. It's John Kerry's campaign now, but it's being powered by Howard Dean's courage and by the grassroots energy that he enflamed.

This is an amazing thing when you consider where Howard Dean started. In a system designed after Jimmy Carter's 1976 win with the sole purpose of thwarting insurgents, Dean for America came within two bad weeks of winning Iowa and New Hampshire, and likely storming on to victory across the country. And the campaign did this even though we began with no major party apparatus, no party money, no support from party leadership, and none of the critical Democratic institutions behind us.

We did this by appealing to the one area politicians have learned to overlook in recent years—the people. We reminded those people of the inherent power that they possess. We planted a seed and a simple idea: If six hundred thousand could turn politics on its head, what could a million do? What could five million do? Ten? If these Internet communities start spilling out into the street, what could we accomplish then?

A quarter of the people who donated to Dean for America were under thirty years of age.[1] This is an amazing number, given that we are living in an age when political involvement among young people is at a historic low, when young people assume they can't make a difference. We taught them that not only *can* they make a difference, it is their birthright as Americans. Twelve of those young people (the Dean Dozen) have already announced their candidacy for public offices. Others have jumped into careers in politics and activism. After the Dean campaign ended, the number of progressive mailing lists and web sites increased as young staffers threw their energies into new political arenas. Among them is a former web team member named Zack Rosen, who is at work on a number of new projects, most notably Progressive Pipes, a web site funded by California tech wizard Andy Rappaport. Progressive Pipes uses the latest in RSS (Really Simple Syndication) technology to help sort through all of the conversations going

---

[1] Anya Kamentz, "Deanie Babies Grow Up," *Nation* (March 2004).

on in progressive blogs, determines what each person wants in his "feed" and provides a running stream of just the information you want. This is the kind of technology that could revolutionize everything, harnessing the power of the Net. Instead of asking for what you want (the Google method), innovations like Zack's cause the Internet to come to you, on your terms. If this is the kind of work that young Dean volunteers and staffers are doing *already,* I can't even begin to imagine what they will accomplish in the next ten, twenty, fifty years.

This is the most important legacy of Howard Dean's campaign for president. From this point on, there will be a new special interest group to reckon with—the American people. And this special interest group has a tool—the Internet—more powerful than broadcast politics, more powerful than all the Pioneers and Rangers and chicken dinners put together. Now politics are no longer the domain of that one-quarter of 1 percent of the wealthiest Americans who give 80 percent of individual political donations. Now twenty-dollar checks from regular Americans can be bundled just like George W. Bush's two thousand-dollar checks. Now Americans stand a real chance against those 631 multimillionaires and billionaires who have been running the country. Now the power is back in our hands.

WALKING OUT THE FRONT DOOR

We had a week to turn things around in New Hampshire. Before the Iowa meltdown, one thing I'd never worried about was New Hampshire, mostly because of the state director, Karen Hicks, who knew the political backroads there like the back of her hand. I'd always said that if we lost there it would be in spite of her, not because of her. Even now with our downward momentum and the "I Have a Scream" speech echoing all over television and on web sites, it looked for a few days like we might actually have a chance in New Hampshire. But even if we did, I knew I wouldn't be around to see it.

Six weeks before the caucuses, Kathy and I had gone to dinner, and she apologized for talking me out of leaving the campaign in October. She could see how the campaign was killing me. I was exhausted. Where I had once gone on the blog to rally the troops with inspiring words, now I went on hoping to be inspired. The cracks in the Dean organization that I had wanted to fix months earlier had become gaping holes, and Howard's lack of trust in me and Kate's open hostility were becoming unbearable. I felt like I was being undercut every day, that I was losing access to, and influence with, my own candidate. That's when Kathy finally told me about the August e-mail in which Kate said she wanted me off the campaign.

As disappointed as I was, it certainly made sense now. We were back on the ocean liner, sailing toward the iceberg, and every time I yanked the wheel to starboard, Kate was going to yank it back to port. And, as I'd noted earlier, this wasn't a power grab. We were two people who genuinely had Howard Dean's best interest at heart. But if one of us didn't let go of the wheel, we'd barrel straight into the iceberg. So that's the moment I decided to let go. Weeks before Iowa, Kathy and I vowed that no matter what happened, we were leaving at the first moment that we could break from the campaign without hurting it: the day after the New Hampshire primary. In the meantime, we decided to give it everything we had, to go hard and try to leave this thing in the best shape possible to continue without us.

I talked to my old friend Mike Ford about it, too. "Bro," I said to him. "If I ever want to jump in the middle of a presidential campaign again—if I ever ever ever ever do that, you come and get me, man. Take me home."

"I will, bro," he said. And then he added, "I tried." It was true. He had warned me about every one of the problems that we still faced. He had done everything he could to keep me from taking this campaign.

The day of the New Hampshire primary, I was in Concord, watching the results. The CNN late exit poll had us right on Kerry's ass—36 to 33, way too close to call. The *LA Times* exit poll had us a point *ahead,* 32 to 31. And suddenly, I found myself imagining what would happen if we somehow won New Hampshire, if we managed to turn this thing around and—

My phone rang. It was Kathy. My wife is an amazing person. A saint. I know it's not easy living with me. Some days *I* can barely do it. But Kathy knows me better than anyone in the world. She was in South Carolina, where she was trying to help build our organization on the ground. "Now listen to me," she said. "No matter how close it is tonight, no matter how well it goes, I want to remind you what we decided six weeks ago. We're leaving, Joe."

I knew she was right. I was fairly sure that the governor and Kate had been planning for weeks to replace me. In their eyes, Dean had been swept up in the momentum of his own populist movement and—once he became the frontrunner—was unable to leave the insurgent firebrand behind and reposition himself as just another moderate Democrat. I blamed the shots we'd taken and our own flawed structure and mistakes for our troubles. Apparently, they blamed me.

So it was obvious for weeks that they wanted to whack me. Roy Neel, a close adviser to Al Gore, had come on board as a senior adviser in late December. Although he wasn't based in Burlington, he was the obvious choice to replace me. They knew it. I knew it. I used to tell Kate and Bob and others in the inner circle, *Look, it's okay. You don't have to sneak around. I know you want someone else. It's fine. Let's just put it on the table.* They acted as if I was crazy. *Joe, we don't know what you're talking about. We're all on the same team here.*

For a few hours on January 27, it looked like the old team might have a few games left. But just about the time we were looking up other words for comeback, the first results started coming in. It turned out that the exit polls in New Hampshire were as wrong as any I've ever seen. It wasn't a dead heat. It was a dead campaign. Howard finished twelve points back, at 26 percent to Kerry's 38. In hindsight, how could he have done any better, when most voters' lasting image of Howard Dean was that looped, out-of-context, overplayed concession speech? Without the wall-to-wall coverage of that hand grenade, I think we might have had a chance in New Hampshire. But that wasn't the hand we were dealt, and in the end, I felt the way I did when Gary Hart left the 1988 race—not surprised, but profoundly disappointed at the cannibalistic fever that our politics can reach, disappointed that we would toss aside a good man for what—yelling?

The campaign staff limped back to Burlington, and the next day, I was called to the law office to see Howard. I had never been to "the law office." Hmm. What could this mean? Howard got right to it. He was bringing Roy Neel in to run the campaign—

"That's great," I said.

He looked surprised.

"No, it's a great idea," I repeated. And it was. This is what I'd been preaching for a year: Put one person in charge and let that person run the frickin' campaign. If it wasn't going to be me—and by now it was clear that it would never be me—then at least get *someone* in there. Roy Neel was a great choice—eight months late. "With all due respect, governor," I said, "what you need to do now is give Roy the checkbook. Let him make the decisions and let him actually manage the campaign. That's the best thing you could do right now. That's the best chance he's got. The best chance you've got."

"But you're staying, right?" Howard said. "You can work alongside—"

"No," I said. "Don't worry, governor. I won't make this hard for you. I'll get out of Roy's way."

He looked stunned. Maybe he was just worried about alienating his Internet supporters, or maybe he genuinely didn't want me to leave, but he explained that he had just assumed that I would stay on with the campaign. He didn't want me to leave; he just wanted Roy to take over the leadership.

"I appreciate that," I said, "but we both know you can't have two captains." I told him that I was going home. Then I told him that he still had a good shot at this. But he had to listen to Roy. Give him the freedom and cooperation that he needed.

It was a friendly, calm meeting—no raised voices, no anger or blame, just Howard's desire to go in another direction, and my desire to go home.

Throughout the campaign, and especially at the end, I occasionally found myself filled with deep emotions that were hard to express. It was a strange sensation, as if I was becoming nostalgic for a moment as I was still experiencing it—some combination of sadness that this was my last presidential campaign, sheer exhaustion over how hard we'd worked, and overwhelming affection for these young people who had battled alongside me, who had given so much to this campaign. These young people who had taught me to believe again.

In November, I'd been overwhelmed with a strong case of this feeling at the JJ Dinner in Des Moines. I had a long, colorful history with the JJ. That day, I watched as streams of Dean supporters climbed off a line of yellow school buses that stretched as far as the eye could see (forty-three buses in all; the streets had to be shut down to accommodate them), wave after wave of Dean supporters in yellow shirts with this written on the back: "How's my organizing? Call 1-866-DEAN-4-US." We had twelve sections of supporters that night, twice as many as Gephardt, Kerry, and Edwards had—*combined*. This is what I had always done. I had gotten people involved in politics, fired them up, got them in shirts and on buses. And this was the best campaign I'd ever been a part of and yet I was always aware that it would be my last one. These were the best people I had ever worked with on a presidential. And they would be the last. I envied these young, talented people, who will, in the next few elections, do amazing things with this new kind of campaigning. But I'd gone as far as I could. That night, at the JJ, Tom Brokaw came up to me and we sat together, watching the speeches and cheers. Tom was retiring from *NBC Nightly News* and this was to be his last presidential campaign, too, and his last JJ. We talked about the old times. You don't want to let go of those moments—but it's like you're just too tired to hold on any longer.

After he informed me that he was bringing in Roy Neel, Howard and I walked back over to the campaign office, and I began packing while he addressed the staff. He talked about the setbacks we'd had and broke it to the staff that we were out of money and that no one was getting paid. And, maybe because he thought that was enough of a blow, or maybe because he still thought he could talk me out of it, he decided not to tell them that I was gone. "No one is leaving," he said.

A young staffer from communications named Garrett Graff came to my office, where I was packing my things, and told me that I should address the room. I didn't want to do it. I was already choking up, but I also knew this would be my last chance, and I needed to look once more at the faces of those people who had done so much over the past year, people who had no political experience when this started, who performed one of the most amazing political miracles I'd ever seen.

When I walked into that room, every face filled me with a mixture of pride and sadness—pride that I had worked alongside such committed, talented people, and sadness that from this point, they would have to go on without me. I looked at the people from the web room—Nicco, Kenn Herman, Zephyr, Mat, Joe Rospars, CarlwithaK Frisch, Joe Drymala—each of whom had stayed up long nights tending what turned out to be the consciousness of this living campaign. *We suck. But for a while there, we were the best team in baseball.* I looked around the traditional campaign staffers, too, talented young people like Paul Blank, Tricia Enright, Tamara Pogue, and Stephanie Schriock, who, at thirty, ran the most successful financial department for any Democratic campaign in history, raising an amazing $50 million, and got almost none of the credit for the most impressive job of fund-raising I've ever seen. There were other people, dozens of them, people who had given every bit of their energy and drive for more than a year.

I had driven them so hard, asked them to do the impossible, and rarely gave out anything resembling a compliment. Once I had chastised them for watching a sunset ("A *sunset?* Do you think Kerry's people are watching the frickin' sunset?") and they'd responded by giving me a photo of a dozen of them gathered at the window, in the glow of the setting sun. Garrett Graff talked about that combination of insecurity and loyalty that I inspired in the staff to a documentary film crew from *CNN Presents:*

> I would follow him off a cliff. And yet, I have never had a positive interaction with him. He's never said a nice word to me, and yet I would do anything for him—he believes in our democracy 100 percent. He believes that it can work. He believes in politics and government as a way to change things and to make peoples' lives better . . .

That's about the nicest—and truest—thing anyone has ever said about me.

That day, I looked out at all those faces waiting to see what I had to say. I just assumed that Howard had told them. "I'm leaving the campaign," I began. There were gasps and a few people began crying. "But I want you all to know—I'm not leaving the movement. I love you all and I'll be there if you ever need anything—"

And then I lost it. The room broke into applause and I made my way back down the hall, while the applause just kept going behind me. I grabbed Kathy. I told her to get her coat. We were leaving. I asked Kristen Morgante to box up all of my personal belongings, leave anything belonging to the campaign, and send the boxes to my home.

Kathy and I hastily got our things together. We worked our way through the crowd, but when we got to the door it turned out that someone had alerted the press to the fact that I was leaving the campaign.[2] The place was crawling with reporters. When Kathy saw them, she turned and

---

[2] CNN reporter Candy Crowley quoted an anonymous source inside the campaign who said that I had "lost it" in my meeting with Howard, which wasn't true and infuriated me because it seemed designed to make it look as if I'd been sacked—one last kick as I went out the door.

ran back into my office, crying, frantically saying we had to go out a back way, or down a fire escape.

I put my hands on her shoulders. I felt strangely calm, relieved this was all over. I'd seen this movie before. I'd seen it in 1988 with Gary Hart. "Baby," I said, "we've got nothing to be ashamed of. We're walking out the front door."

And we did. We went back to the house where we were staying and for the first time in a year, I slept all night. And the next morning we got up, and without packing a thing, climbed in our car and drove fourteen hours, through the Northeast, down the Eastern Seaboard, past Washington, D.C., across the Chesapeake Bay and onto a finger of the eastern shore of Maryland. We turned down our unpaved driveway and, finally, after fourteen months, we were home.

PART III

SEIZING POWER IN THE INTERNET AGE

11

THE BEGINNING—1956
Google, Napster, and the Disney Dweeb

THIS IS THE BEGINNING

FROM HERE, THE world changes faster than you can imagine. This is where it gets good, where it gets thrilling, frightening, inevitable.

Best of all, this is a fresh start. We get another shot at this.

It's 1956 again and we just got the box in the house.

We are in the same stage of the Internet revolution that the television was at in 1956, when the percentage of homes with a TV passed 75 percent. In 1956, it was 45 million people watching *The Wizard of Oz* on TV, Elvis appearing on Ed Sullivan, and the networks covering the Democratic National Convention for the first time. Today, 75 percent of Americans are online, eBay draws forty-five million bidders a year to its auctions, hundreds of million of songs are downloaded from music web sites, and the Dean campaign has shown how a candidate could ride the Net from obscurity to the verge of a party's presidential nomination.

Clearly, we are on the cusp of a fading era and a new age, just as we were in 1956. The Internet is like an eight-year-old child, growing in leaps and bounds, giddy with possibilities. Look around you at the eight-year-olds. Try to imagine what the world will look like when those eight-year-olds are forty. Or fifty. Those kids who operate a mouse like an extension of their hand—is there any question what their dominant technology will be? Remember what the world was like when you were eight, then look at all that has changed in fifty years.

In 1956, there were three black-and-white channels (*Gunsmoke* and *The Honeymooners* were starting). TV programming ended at midnight and NBC was about to revolutionize late-night test patterns with a band of peacock-tail feathers that a fraction of TVs showed in muted color. Every night we crowded in front of this box for entertainment and a bit of news about the scourge of Communism.

Who would have guessed that the huge console in the corner with three black and white channels would become a flat screen hanging on the wall where we could watch five hundred channels in wide-screen letterbox satellite pay-per-view high-definition color? Who would have believed that we could watch movies, record programs, take video of our family and watch it on the television? Who would have thought we could record any program we want and watch it whenever we want, without commercials? Who would've imagined that we would watch music on TV, attend church on TV, that our elections would be contested on TV, that we would watch acts of terrorism on TV, that our wars would be waged on TV? Who could have predicted the myriad ways our culture would be transformed by that box, the way our lives would revolve around it?

This is the same effect the Internet is having on our lives today. Some brilliant minds have been at work for the past decade studying this. These are the pioneers of a new world: writers, thinkers, and innovators exploring the boundaries and creating maps the rest of us can follow. Among them are several visionaries whose work provided a jumping off point for Howard Dean's presidential campaign, including:

- David Weinberger, the prescient author of *Small Pieces Loosely Joined,* a lucid exploration of the way communities form on the Internet and the idea that "we are the true 'small pieces' of the Net, and we are loosely joining ourselves in ways that we're still inventing."
- Lawrence Lessig, *the* expert on intellectual freedom and the Internet, whose vital book, *The Future of Ideas,* envisions a world in which a thoughtful balance between the offline tradition of private property and the Internet's sense of a knowledge commons sparks a renaissance of innovation.
- Howard Rheingold, whose seminal books *The Virtual Community* and *Smart Mobs* first proposed that computers and then cell phones and PDAs would be technologies that foster community building, transform

cultures, and "change the way people meet, mate, work, fight, buy, sell, govern and create."

• Doc Searles, Chris Locke, and Rick Levine, who, along with Weinberger, collaborated on *The Clue Train Manifesto,* a book that one reviewer called an "obituary for business-as-usual" in America. The book proposed that "markets are conversations" and that people are drawn to the Internet because it offers "in sharp contrast to the alienation wrought by homogenized broadcast media, sterilized mass 'culture,' and the enforced anonymity of bureaucratic organizations, the Internet connected people to each other and provided a space in which the human voice would be rapidly rediscovered."

These pioneers and others have been out in the wilderness for years telling us what to expect, sometimes painfully ahead of their time. When the Dean campaign attempted to harness the power of Rheingold's *Smart Mobs*—people organizing via wireless text messaging—we created the largest network in the country, about 5,000 people. But the technology was still a little raw for political organizing in the year 2003. In the same way that Howard Dean's Internet support was roughly ten times that of John McCain's four years earlier, when Rheingold's *Smart Mobs* make a similar tenfold jump in influence, the candidate who figures out organizing through text messaging could win the presidential election.

In that way, Dean for America is a sneak preview of coming attractions—the interplay between these new technologies and our old institutions. The end result will be massive communities completely redefining our politics, our commerce, our government, and the entire public fabric of our culture.

For ten years, it has been the dawn of this movement. Now it has arrived. For years, we've seen the Internet as a revolution in business or in culture. But what we are seeing—at its core—is a political phenomenon, a *democratic movement* that proceeds from our civic lives and naturally spills over into the music we hear, the clothes we buy, the causes we support.

What this democratizing means for American businesses is that they have a choice now: Embrace a more open, responsive era, or make a doomed stand against progress. Cling to the old top-down structure, or radically reinvent the way you do business. The companies that survive and thrive in this new age will be the ones that respond to the three demands that this movement will make of corporate America:

1. We want businesses that empower consumers.
2. We want corporations to be responsive to shareholders.
3. We will only tolerate companies that are good corporate citizens.

This is the real choice for American companies. Will they willingly invite their customers inside or push them away? Will they listen to small groups of shareholders or wait until they band together and take over the boardroom? Will they stop polluting and start taking care of their employees, or will they wait until we force them to do it.

CUMULATIVE INTELLIGENCE

In the spring of 2004, the biggest business story in the country (at times, it seemed, the only business story) was the announcement of an initial public offering for Google, which was expected to raise some $2.7 billion in stock sales when the company went public sometime in the summer.

Google is a phenomenal business, to be sure. Just five years after two Stanford PhDs designed what would become the world's favorite search engine, it took in a billion dollars in gross receipts and more than $100 million in profits. But Google is more than just its bottom line. There are plenty of other search engines, run by some of the biggest computer and technology companies in the world. Most of them do the same thing.

So how has Google outperformed them all?

I think the company's motto says it all: "Don't Be Evil."

While so many other companies sell every square inch of their home page for advertisements and cover them with graphics, Google's home page is nearly empty, to allow for slower dial-up modems and faster service. Google does show you ads—but only a few, about topics you inquire about, and the ads are clearly labeled, off to the side and as unobtrusive as possible, unlike other search engines, which slide paid links into your search results.

As Google is preparing to fire up its new free e-mail service, Gmail, it is following its own motto, offering people a gigabyte of memory, five hundred times the amount of memory that MSN's Hotmail offers, and the ability to Google your own messages for information you might need.

Other features, like automatic spam removal and the length of account life, hint at why Google is so successful:

> [O]nce thousands of people begin to use the Report Spam button, Google plans to harness the cumulative intelligence of its customers to refine its spam filters in innovative ways.
>
> Finally, Google promises that it won't shut down your account until you go nine months without using it. (Hotmail and Yahoo delete all your mail and recycle your address after only 30 days.) Now that's not being evil.[1]

This is more than a commitment to good customer service. Google's spam-removal system would "harness the cumulative intelligence of its customers." This is the heart of "grid computing," the oldest idea on the Internet, the basis of the entire thing, the idea that a million computers are more powerful and more secure than having all that information on one large supercomputer. This is the innovative idea behind grid.org, where anyone can download a screensaver that puts your computer to work on a common problem while you're not on it, which is why millions of personal computers are working right now on a cure for cancer. This *is* the Internet, the simple and radically democratic idea that a million computers are always better than one, a million people are always smarter than the biggest corporation, that collective power is our greatest wealth.

Just as we, in the Dean for America campaign, assumed that we might not have the twenty smartest people on the planet in our office, and that Howard's 660,000 supporters might not all be morons, Google realizes that the best ideas about what its customers really want might just come from . . . *its customers.*

This sounds so obvious as to be idiotic, but it's amazing how many American companies proceed as if the oracles live in their boardrooms, emerging from their glass-walled caves to make proclamations about the world. And when these old-world businesses *do* decide to seek input from customers, it packs them into a fish-tank focus group and studies them like lab rats. *What happens if you give the rats this much tobacco? What happens if you show them this commercial?* Is the point of these focus groups what people want? Or what the company can get away with?

---

[1] David Pogue, "Google Mail: Virtue Lies in the in Box," *New York Times* (May 13, 2004).

That's going to be over soon. What's the point of a focus group of forty when you can ask all of your customers what they want and they can tell you?

The companies that will thrive—*the companies that are thriving*—are those that embrace the bottom-up nature of the Internet culture. The companies that make the turn will be those that figure out how to empower their customers to have a say in the products they buy and use. This will do two things for those businesses:

- Customers will be a part of the process, and will naturally form a community around the product or service, the way people feel about eBay or Google, or around their favorite blog. These people aren't just users, any more than you are *using* your house. They *live* there. They pick up garbage on the street (spam and trolls). They watch each other's homes and keep an eye out for criminals (viruses and online scams).
- Your company will get better. This happened to us so many times on the Dean for America campaign I lost track. When people know they are being heard, they will speak up, and when they speak up, they will offer ideas that never occurred to you or your $60 million-a-year marketing team or your billionaire board of directors.

I'll give you an example of how this could work. The great thing about being a media consultant is that people sometimes pay me for this stuff, but in the spirit of Open Source, this one is free. Ford Motor Company: The venerable automaker has a slick, nifty-looking web site. And it's like watching the most boring television commercial you ever saw. It hardly ever changes. There are pictures of cars, some words, a place you can look at press releases and corporate information. (Yeah, like that's going to compete with online porn.)

But what if Ford did this: Announce that it's designing a new Mustang and that it wants *all those loyal drivers who ever owned a Mustang* to help decide what it should look like. Say, pick a new color. People could go to the web site, run through photos of the car in a whole palette of colors and vote for the one they like best. Maybe they can even vote on a design element, whether it has a fastback or not. Now, what happens when they roll that new Mustang out on the showroom floor? You will have created a community that has something *invested* in this car and, best of all . . . *it will actually be a car the community wants.* They'll buy it for two reasons: It has features they asked for, and they already have ownership in it. A group that already exists—intensely loyal and nostalgic Mustang drivers—will form a community and have a home, the Ford web site, where they can compare mileage, share service tips, and offer suggestions for next year's model.

Compare this to the way cars are unveiled now, like top-secret weapons. Then, once the manufacturer has rolled out its new car, it has to compete for the attention of the automotive media (*Car and Driver,* etc.). And even if the automotive media gives the green light, the company still must buy hundreds of millions of dollars in expensive, scattershot advertising in the hopes that they'll accidentally and randomly hit people who are watching *Survivor* and have been thinking about getting a Ford. Why not have the people who drive Fords come to you? And when they get there, why not let them *do something?*

I suggested a similar idea when I spoke in front of an advertising group that has a prominent beer producer as a client. I asked how many times this beer company had launched a new kind of beer. Beer Light, Beer Dry, Beer Ice, Beer Clear. . . . Then I asked if they'd ever let the people who actually drink the stuff choose the name of the new beer. *"If it's going to be your beer, shouldn't you name it? Go to Beer.com and name America's beer."* The ad people looked around at each other and nodding slowly as the simple logic of the idea washed over them. Keep your eyes open. If it hasn't already happened by the time you read this, it will. And the companies that figure this out first will be the ones that survive the revolution. The democratic movement I'm talking about empowers consumers as well as citizens.

Of course, it's a big step between letting someone choose the color of a car or the name of a beer, and truly empowering people.

That's why a great company like Microsoft could find itself in trouble. The five-hundred-ton gorilla could eventually lose its grip on the computer operating system and software business unless the company radically changes the way it does business. Today, Microsoft essentially puts two thousand really smart software engineers in a dark room and tells them not to come out until they have the next version of Windows. The code key for the software they're writing is locked in a vault. Forever. It's gold. If that key ever got out, Microsoft couldn't make money.

Here's the problem: Those two thousand smart engineers bust their asses, write millions and millions of lines of code and they make a few mistakes—typos, mostly—and each of those mistakes is a hole. And those two thousand engineers, as smart as they are, can't possibly be expected to find all the holes. They're too close to it. So the second they release Windows '09, this whole group of people whose reason for living is to find the holes in computer programs will swarm in. When a virus takes down your computer, that's what's happening. Some hacker is slipping in through a hole that Microsoft missed.

Compare Microsoft's ethic to that of the old Internet (and the ARPAnet), which coalesced around a free source code. At MIT in the 1980s, a guy named Richard Stallman started the Free Software Foundation in the spirit of the old Net, writing a new operating system that would be licensed so that it remained open to the public, essentially free to use, to copy, to update, to improve—so long as no one tried to copyright the improved version. But Stallman's system (called GNU) lacked one essential component: the kernel, the central code key, like the one that Microsoft keeps locked in a vault somewhere.[2]

In 1991, Linus Torvalds, a student at the University of Helsinki, began work on the kernel to this system. Torvalds took Stallman's free software one step further by posting his code on the Internet and *inviting* people to improve it, asking for help from all of the other computer users out there who were concerned about Microsoft's cornering of technology that had once been shared by its users. The key he came up with is now called Linux, and it's the fastest growing operating system in the world.

Unlike Microsoft, when Linux users found holes they could go get the key, unlock the code, fix the typo, and put the key back. All of a sudden, all over the Internet, millions of people started writing in Linux, millions of people looking over the shoulder of Torvalds, instead of just two thousand. Instead of having just the hackers look for holes, Linux had the good guys looking, too, and they usually fixed the holes before the bad guys could get in.

So Linux is—for the most part—free of holes. There has never been a serious Linux hack. It's more stable, it's free, and it's out there. Anyone

---

[2] Rheingold, *Smart Mobs*.

can download it and begin using it. Microsoft, of course, hates the fact that a more stable operating system exists that doesn't have the potential for viruses and that is, worst of all, free.

So why isn't Linux on 90 percent of America's computers instead of Windows? Microsoft got a big head start, and in the computer world, the *first mover* is still the winner in most cases. And all of that software written for Windows guaranteed that Bill Gates's operating system would remain on top.

But Linux is growing quickly. New software is becoming available all the time. Microsoft is, obviously, a great company. But the rules are changing. Take your computer for example. What's the most expensive thing in the whole damn box? It's not the Intel Pentium chip. It's not the screen. In a five hundred dollar computer, Microsoft is pocketing between one hundred and one hundred and fifty bucks just for the operating system. And that's the break the computer manufacturers get. Buy Windows for your own machine and you easily pay twice that.

So if you're a manufacturer, and you can switch to a free operating system and knock a hundred bucks off that five hundred dollar box, wouldn't you do that? All things being equal, do you want to pay five hundred dollars for a computer? Or four hundred? This is why if Linux ever gets to a point where the available software comes close to Microsoft, Windows will become an exhibit in some museum.

But the eventual success of Linux is based on more than just money or competitiveness or even superior technology. It's about the culture. It's the same reason Napster and Google and eBay have thrived. It's because there is a moral kernel to this revolution, a code key that is out there for any company to pick up, use, understand, improve, and put back out there for other companies to use. And it is this:

Don't be evil.

Businesses now have to ask themselves: Which of these things makes us a better company in the long term? Not necessarily a more profitable company in the short term. These are not just markets anymore. They're communities. And we're not just consumers. We're citizens again. We're looking for the companies, politicians, and institutions that will build the best communities. These communities will not be geographic, but will be constructed of people with similar interest, aesthetics, and beliefs. We'll remake the world the way we want it to look. This is what we learned at Dean for America. On the Internet, everything is a community. Every site is a place. As Gary Wolf wrote in *Wired*:

> Joi Ito, founder of Neoteny, a venture firm, and former chair of Infos-eek Japan, has joined a group of technologists advising Dean . . . I contact him to ask if he thinks there's a difference between an emergent leader and an old-fashioned political opportunist. What does it take to lead a smart mob? Ito e-mails back an odd metaphor: "You're not a leader, you're a place. You're like a park or a garden. If it's comfortable and cool, people are attracted. Deanspace is not really about Dean. It's about us."

In the end, that's why I believe Linux will eventually win, because no matter how many software engineers Microsoft puts in that dark room, it's never going to look like a garden. Their model makes great software, but it doesn't create beautiful parks. Microsoft's community is like an old rail-road town. Meanwhile, the innovative, free community that has sprung up around Linux will only grow.

Obviously, this will take some time. We are still at the end of the old TV-dominated world, the beginning of the age of the Internet. Today, not every business can follow the Linux example and go totally open-source. But I believe that finding that balance between decency and profitability will determine which businesses survive the revolution. You can think of it as the Don't-Be-Evil test. This will be the challenge for businesses in the Internet Age, going beyond the symbolic involvement of their customers to truly empowering them.

AMERICAN IDOLS

His name was Shawn Fanning, and he was a nineteen-year-old Pink Floyd fan living in Cape Cod. But because he used to wear his hair in nappy curls, his friends all called him Napster. And all by himself, this kid revo-lutionized the recording industry . . . and very nearly destroyed it.[3]

---

[3] David Kushner, "Napster Terrorizes Music Biz," *Rolling Stone* (April 27, 2000).

As a freshman studying computer science at Northwestern University, Fanning wrote some software so he could share MP3 music files with his friends. Soon, with his program, Napster, people could go online and peruse the music libraries of their friends or people with similar musical tastes (or people with no musical tastes, for that matter). Within a year of the time he got the idea, Fanning's Napster was an Internet phenomenon, a worldwide digital-music co-op.

The technological key to a program like Napster, the particular method of community computing, is called "peer-to-peer" or P2P. By the time Fanning wrote his program, the concept of P2P had been used a million different ways for years on the Internet, connecting dozens of computers to do huge jobs, or just creating networks of similar-minded people. It's what Rheingold calls a "technological windfall":

> People don't just participate in P2P—they *believe* in it . . . the unique human power and pleasure that comes from doing something that enriches everyone, a game where nobody has to lose for everyone to win.[4]

And when it was confronted with this vision of the future, what did the music industry do? They did everything they could to kill it. The Recording Industry Association of America brought lawsuits against Napster for copyright infringement—for distributing music to people who hadn't paid for it. Soon, Napster was out of the free music distribution business. This was around the time the Internet bubble was bursting and the techno-naysayers had a field day with this one.

But pick up the story four years later and you see that Fanning's revolution proceeded with the inevitability of all democratic movements. Napster, of course, is back, one of dozens of companies around the world that allows people to download music (for a price), a song or an album at a time, onto MP3 players or computers. The biggest is that old computer innovator Apple, which sold three million iPod players in the past year and whose customers downloaded *seventy million songs* in the first year of iTunes, its version of Napster. Each song costs ninety-nine cents, albums are ten bucks, and the iTunes catalog features some seven hundred thousand choices. Apple has even tried to bring back at least some of the P2P spirit of Napster with

---

[4] Rheingold, *Smart Mobs*.

iMixLists, music that can be e-mailed to friends and family in free thirty-second song previews.

This is one of the fastest growing segments of the economy, with music distributors and players flooding the market. But make no mistake about it. No matter how much money Apple or Microsoft makes, this remains Shawn Fanning's revolution.

And the ridiculous thing is this: The recording industry could have done this themselves at any time. They could be making the money that iTunes is making. They could have profited from the goodwill and the sense of community. Look at the VHS industry. For years, Hollywood fought videotape recordings, fearful that it would kill movie box office business. Now it's the most profitable segment of their business. The recording industry could have figured that out. Instead, they tried to kill it, and they were run over. The music industry bigwigs chose to fight and the nineteen-year-olds kicked their asses and totally changed the way we listen to music.

Entertainment is the natural leading edge in this movement (remember how the *Blair Witch Project* was marketed a few years ago). Books were the first mover of first movers, when Amazon.com established a profitable online business while the rest of the world was still studying whether it was possible. But the book phenomenon that best illustrates what's in store for American business is Book Crossing, a creative book-sharing web site that has grown from 16,000 members on Meetup.com a year ago to more than 37,000 in the spring of 2004 (and tens of thousands more who visit its web site—bookcrossing.com). They meet on the second Tuesday of each month, give their opinions about books (on the web site this is done in short journal entries) and then "release the books into the wild." Books are left on park benches, on the shelves of laundromats, and on street corners. Each book carries the history of its journey (and the reviews it's gathered along the way) as well as some personal information about the people who read it. People are connecting with literature and with each other, and Book Crossing is one of the most vibrant communities on the web, an organic movement growing on its own. The question is whether publishing will find a way to embrace this community of passionate readers (Readers!) or whether they'll put their collective heads in the sand the way the recording industry did and wait for the nineteen-year-olds to break down the walls.

And we can't forget television.

The emergent technology of 1956 is embracing the Internet revolution as fast as you can say "reality show." That's all a reality show is, television trying to compete with the variety, authenticity, and instant feedback of the Internet. It began with polling the audience on *Who Wants to Be a Millionaire?* and it's evolved to polling the whole world on *American Idol* (which is the Internet on TV). We watch real people sing on TV and then we call in on our cell phones or we text message the people we want to advance to the next week. But the most important thing about *American Idol* is not how it's done, but what it reveals. The producers clearly thought that when they gave Americans a choice of "idols," we would choose the bland pop music dolls that dominate radio—the crap *they've been feeding us for years*—and so they overstocked the pool with Justin Timberlakes and Britney Spears.

But when Americans get the choice—when bad singers in tight clothes aren't simply shoved down their throats, but are put alongside actual *choices*—they constantly surprise the producers and the celebrity judges. They go for gospel singers and torch singers and big band singers. They vote for fat people and geeky people and ugly people. They go for people like themselves. They choose a stunning variety of musical genres, singers and songs that would never occur to record producers. This is the most important thing that any business can learn from the first wave of this revolution and its impact on entertainment:

We want the power to choose.

That's the secret of TiVo. For years, we were slaves to our televisions. We watched what they put on when they wanted us to watch it and we stared at whatever commercials they put on in between the shows. Now we choose what we want to watch when we want to watch it. And I hate to break it to you . . . we're not watching the frickin' commercials anymore. The rebellion is well underway on TV. From three channels of tightly controlled and scheduled programming and commercials to hundreds of channels that we schedule, that we control . . . we are finally gaining control of that monster in our living room.

In every industry, in every segment of our economy, the power is shifting over to us. If you're working for an American company, you get to decide whether you want to be the recording executive who thought he'd beaten Napster, or the one off trying to figure out how to make money with the iPod.

You get to decide if you're going to keep throwing money at television advertising, a one-to-many medium that becomes more expensive and less effective with each passing day, or if you're going to do what eBay did and build a community around your company. Because if you're not committed to building the Kodak community or the Chrysler community, then you're just putting in your time until some first mover in your industry does it and takes you out. Then you'll say you never saw it coming, just like the recording industry never saw it coming, just like the movie industry won't see it coming when kids start using digital cameras to produce two-hour movies and distribute the films themselves using broadband.

I promised myself I wouldn't do this: try to give the reader what so many other books promise—the Seven Steps to Interconnectivity or the Nine Keys to Unlocking the Secrets of the Net or some such crap. The truth is that it's impossible to break the power of the Internet down into a recipe of easy-to-follow steps. However, there are several basic principles that companies can choose to embrace before it's too late. Take 'em or leave 'em, here they are:

The Seven Inviolable, Irrefutable, Ingenious Things Your Business or Institution or Candidate Can Do in the Age of the Internet That Might Keep You from Getting Your Ass Kicked But Then Again Might Not

1. *Be first.* There is very little about the Internet that is proprietary. I could start an online bookstore tomorrow and do everything Amazon does. And you know what? Amazon would still beat me like a dirty rug. It's about more than branding. The first car company to let people pick the colors, the first beer to let people design the label, the first candidate to embrace people on the Net—the first *everything* has a head start building a community. Go now. (Rule 1a: If you're not going to be the first mover, you'd better be a hell of a lot better.)

2. *Keep it moving.* Do not be static. The Internet is a liquid medium. It's amazing how many companies spend $100 million on TV advertising while their $64,000-a-year "web division" consists of the CEO's twenty-two-year-old Nintendo geek nephew updating the web site with a new press release once a month. Don't let your web site be wallpaper. Your Internet presence should be an organic, flowing, daily

dialogue with your customers, back and forth. If you aren't regularly e-mailing customers, if you aren't responding to their e-mails, if you don't have a blog, if you're not using your web site to engage the people around you . . . then you are wasting your time on the Net.

3. *Use an authentic voice.* The blogging expert Dave Winer calls it the essential element of web writing: "the unedited voice of a person." The Internet is not the place for safe, vetted corporate communication. We're not morons. When we get an e-mail from the president of the company, we know it wasn't really written by him. People would rather get a real e-mail from a real guy in the real mailroom than a phony one from the CEO (who we know is vacationing on his yacht anyway). Sacrifice some of the slickness of your web site for the real, sometimes messy quality of the best blogs. And no more autoresponses. Have real people write real stuff.

4. *Tell the truth.* The Internet has an inherent transparency. A strong Internet presence is a way to open the doors of the company. But if you invite people in, you'd better be prepared to have them look in the medicine cabinet. So don't hide anything. Tell them what you want. Don't manipulate. Put *what you want* up high. Put it on the first page of your web site, at the top of the e-mails.

5. *Build a community.* Create a commons, a town square, a place where people can come together to talk about their Ford Mustangs, or their Kodak cameras. If you are running the Kodak web site and you don't have an online photo gallery for the people who buy your digital cameras, or an online photo contest . . . then you should give up now. Because someone else is going to do it. *Get people involved!* This is not top-down, one-to-many anymore. The Internet is side-to-side, up-and-down, many-to-many. Use it that way. It's the dialogue, stupid.

6. *Cede control.* Once you invite the people in, they're going to want to do more. I know this violates everything they taught you in school, but you have to let go of the old command-and-control style of business. Let the edges blur between customer and company. And remember: *We are smarter than you are.* If you let us choose the color of Mustang, you'd better be prepared to produce some squash-colored cars.

7. *Believe again.* The days of condescension toward customers and citizens are over. Have some faith in the American people again. Democracy is based on the principle that if we give the citizens control over

their common future, they will choose the best path. The same is true of consumers.

You can adapt, now, or you can wait for the eager, hungry, wired nineteen-year-olds—and the rest of us acting with the energy and fearlessness of nineteen-year-olds, whether we're nineteen, thirty-nine, fifty-nine, or eighty-nine. We will burn with the desire to overthrow everything. You can wait for us, or you can change your way of thinking, embrace this democratizing power. You can build communities and contribute to a new way of business, government, life . . . or you can wait for us to come over the walls.

And when we get there, what will we want? Good stuff—good clothes and good cars and good music and good public policy. Good corporate behavior. We'll want you to stop polluting and stop creating debt that our children will have to pay for. We'll want you to listen, to make good communities where we can shop and play and date and listen to music. We'll want you to not be evil.

DELUSIONS OF ADEQUACY

Jim Hill was in his sweat pants, taking a nap on his couch at home in New Hampshire when the phone rang. It was Roy Disney, the nephew of Walt Disney. Once a board member of the company his uncle had founded, Roy was involved in a fight over the future of the company and was desperately trying to get rid of Disney's chairman, Michael Eisner, whom he believed was ruining the company.[5]

So what did a powerful millionaire, heir to one of the great companies in America, want with a guy in New Hampshire wearing sweat pants, a guy who describes himself as "a Disney dweeb" and "the weenie in the woods"?

He wanted Jim Hill's help.

Hill is a fan of all things Disney, part of a huge community of people who trade information about everything from price per earnings ratios to why the Teacups ride at Disneyland has gotten so slow. Jim runs a web site

---

[5] This Disney example comes mainly from two sources: Sarah McBride, "Mice That Roar," *Wall Street Journal* (March 2, 2004); and Deborah Solomon and Joann S. Lublin, "Voting Rights: Democracy Looks for an Opening," *Wall Street Journal,* March 22, 2004.

called JimHillMedia.com, a blend of news, gossip, insight, product, and movie reviews, laced with Jim Hill's self-deprecating humor:

> Regular readers of JimHillMedia.com will know that—in spite of this web site's rather self-aggrandizing name—I don't really have delusions of grandeur. Most mornings, I'm lucky if I can actually pull off delusions of adequacy.[6]

Roy Disney called Jim to enlist his help in a campaign to wrestle the company away from Eisner. As the Disney shareholder meeting approached, Jim Hill's "little, teeny" web site was attracting 1.5 million hits a day from other Disney fanatics. In a fight over control of the boardroom, Roy Disney realized that he needed the help of small investors who would never come to a shareholder's meeting. The beauty of it was, with Jim Hill's help, they wouldn't even need to leave their living rooms.

As the *Wall Street Journal* reported:

> Disney is a rare company, attracting thousands of people who care passionately about it. They visit Disney parks, watch Disney movies, dress in Disney costumes and spend countless hours thinking about the company. Dozens of Web sites cater to this group, which obsesses about every detail of the company. . . .
> But unlike Trekkies or Deadheads, who are into Star Trek and the Grateful Dead, hardcore Disney fans own stock in the object of their devotion. Those stakes may just be a few shares apiece, but multiplied by thousands of fans across the country, they help make Disney one of the most widely held stocks on Main Street.[7]

In March 2004, 43 percent of the shareholders who voted rejected Michael Eisner, and he was stripped of his title of Chief Executive Officer.

Disney may be a rare company, but what happened to it in 2004 is going to start happening in boardrooms across the country if CEOs don't wake up to the revolution around them. At the same time the Internet is making it possible for small investors to band together into more powerful voting blocs, the agency that governs corporate governance is looking for

---

[6] JimHillMedia.com (May 10, 2004).
[7] Sarah McBride, "Mice That Roar," *Wall Street Journal*.

ways to expand democracy among shareholders. In the spring of 2004, with people outraged about the weekly flood of corporate scandals, the Securities and Exchange Commission (SEC) began weighing rules that would allow a smaller percentage of shareholders to nominate directors to the boards of companies—in groups as small as 5 percent.

The SEC first began considering the idea that investors could nominate board members in the wake of the Great Depression. But what followed was a string of rules that actually diminished democracy in the board room:

> Corporations responded that shareholders would make foolish decisions and elect irresponsible directors. Milton V. Freeman, an SEC attorney who wrote the plan, was labeled a Communist by some members of Congress. The SEC never implemented the proposal.
>
> Soon, the SEC sought to bar proposals of a "political, social or economic nature." In 1951, the SEC rejected a request by Greyhound Corp. shareholders to submit a proposal asking management to end its segregated seating policy in the South.
>
> Three years later, the agency formally clipped shareholder rights by prohibiting investors from making proposals related to a company's "ordinary business." To this day, companies use that rule to exclude all sorts of proposals, such as boosting the cost-of-living adjustments for retirees.[8]

While the corporations were fighting to keep segregation and low pensions for its retirees, there were some activist institutions, like pension funds, battling to get boards of directors to do what's best for employees, investors, and the world at large. These institutions have long owned huge chunks of major companies—as much as 10 or 15 percent—but it has never been enough to effect real change. But with the SEC considering new rules (assuming corporate lobbyists don't take the teeth out of them), these pension funds and other institutions will join with regular investors using the Internet to seize tremendous power. Good luck outsourcing work to third-world countries when we're in charge, or denying basic benefits to employees (*Attention: Wal-Mart shoppers*) or paying your CEO $200 million a year. We're only at the beginning, but the bottom-up impact of having *us* in the boardroom will result in a whole new world of corporate governance and responsibility.

---

[8] Lublin, "Voting Rights: Democracy Looks for an Opening," *Wall Street Journal*.

Maybe your company has embraced the Internet as a marketing tool, or as a way to get feedback, but stopped there. It's a start, but it's not even close to being enough. This revolution is about more than just consumers. It's about citizens, too. And, in the end, the company itself is just another citizen.

Companies like Google understand this, which is one reason it's invited the masses to its Initial Public Offering (IPO). In the old shady brokerage days of IPOs (yesterday), investment bankers, company insiders, and Wall Street bigs got the first crack at shares of a new company; they bought up all the shares, started a buzz and then turned around and sold half their shares two days later at an obscene profit. This was a layer of grease the small investor never got to see. Google has rejected the old corrupt IPO system and is offering its shares in a public auction. Anyone can register to buy the shares. In fact, Google's prospectus reads like a manifesto against the old system, complete with the kinds of consumer warnings that you usually find on a pack of cigarettes: "Our brand could be tarnished, and users and investors could become frustrated with us, potentially decreasing their use of our products and services."

Giving everyone a shot at profiting from your company? Warning investors that it could all fall apart? How egalitarian. How un-evil. This is the world that companies find themselves in now. This is the world that Americans are beginning to demand, the world that companies will have to embrace if they intend to survive. The measure of a great company will be the way it builds great communities—not how much it saved by screwing its pensioners. The only question for companies wishing to do business with us will be this: Are you building a place where we want to live?

Companies can do this in a billion ways, from letting us design our own Mustang to polling shareholders before outsourcing jobs to Malaysia. But the real first movers will be the companies that take it one step further. It's not just: Don't Be Evil. The real First Movers will be the ones that figure out a way to: Be Good. Take on the same responsibility of building a better America that we expect of our citizens.

United Parcel Service is another great, iconic American company, who is unparalleled at whisking packages around the globe. People who use UPS generally don't use Federal Express or Airborne Express or any other delivery system. They go online to check the progress of their shipments. They already exist in a kind of loose community. Here's what UPS should do. A month before Christmas, UPS should announce that in the spirit of the holiday, the company is going to devote one day, let's say December 20, to picking up special packages: Food. Toys.

Ads will run on TV and e-mails will go out on the Internet: "This holiday season, UPS wants to give something back to the nation that supports it. Please help us feed those people who don't have food, and help us give toys to those kids who don't have anything to play with. Just put together a package, go sign up on our web site, and we'll do the rest."

"UPS and You—showing in one day what America can do."

People by the millions will go online and sign up to have a UPS driver pick up nonperishable food and toys for kids. The biggest obstacle to charity is that people just don't get around to it. How long does the bag of clothing sit there before you donate it? But if the UPS driver comes right to your door, you will be ready. The tons of food that the company gathers that day can go to food banks. The toys can go to Toys for Tots and other charities.

What will UPS have done? It will have reinforced the importance of its own community. It will have created a project for that community, the same way real-world neighbors might gather to clean a park or to read to kids. The members of that community will feel better about themselves and better about UPS. And they will be loyal to a company that makes them feel good about themselves and is committed to a better world. And people who have never used UPS will have paid a visit to the community. People will have experienced the ease with which a friendly UPS driver comes to the door and effortlessly picks up a package. And best of all, *you will have fed the poor. You will have given toys to kids who would have gone without them at Christmas.*

We waited two hundred and thirty years for the promise of meaningful, democratic power in this country. We're ready. We're ready to demand better from the corporations, from the political parties, from the government, from our schools, from ourselves. We are a nation connected—not only by a mesh of wires—but by a phrase that the founding fathers once used to define us, a phrase that we allowed to disappear from the national lexicon, a phrase that we're ready to demand from the institutions that are supposed to serve us:

The common good.

12

THE AGE OF THE INTERNET

A Little Rebellion, Trent Lott, and Jefferson's Revenge

A FEW WEEKS after I left the campaign, Howard Dean dropped out of the race for president. In spite of his new management team's attempts to move him to the center, his campaign never recovered from broadcast politics' overkill coverage of his Iowa concession speech, or from the low blows that he took every day during December and January. He returned to Vermont with a stronger voice in the national debate.

Kathy, Kasey, and I settled back into our rambling, listing house on Cummings Creek Farm on the Chesapeake Bay of Maryland. I had spent the equivalent of one week there—seven nights total—during the campaign, so as we drove up the long driveway, I looked forward to catching up on that year of sleep. I was exhausted but I couldn't rest.

One day during the Dean campaign, I had spoken to my daughter Christine's high school class in Evanston, Illinois, about the role that young people can play in politics. Hoping to warn them about not becoming disillusioned adults, I asked a roomful of high school seniors how many of them thought they could contribute to making a better world. I figured they would all raise their hands.

But after I asked the question, not a single hand went up. In a class full of seventeen- and eighteen-year-olds, not one person believed that he or she would be able to make a difference.

If I had needed any more proof that change is long overdue in this nation, those blank stares in my daughter's class provided it.

In the weeks following the campaign, I talked to other Dean staffers. It was remarkable how many felt the same way that I did. We had come through this thing together and we'd come close enough to know that change was possible. And I think we all felt as if something was left undone. In February 2004, a few days after Dean left the race, about forty-five staffers and volunteers from the campaign gathered for a weekend at the farm. We laughed and reminisced as they presented me with some mementos from the campaign, including the framed red bat from Bryant Park. During the day, these amazing young people helped clear away brush deposited by Hurricane Isabel, which had covered our land with debris the winter before. Then, on the last day of the weekend, we had a big breakfast and talked about the campaign, about the momentum we'd built and the enduring promise of progressive politics on the Internet.

We all agreed there was more we could do. For starters, George W. Bush was still the president, in the process of raising some $200 million to buy another tight election. But more than that, I think we all believed that even though our candidate had gone down in defeat, the *campaign* he'd led was still very much alive in those 600,000 people who were still out there. And we all believed that there were countless more like them, millions of people looking for ways to connect to one another in a community bound by a desire for economic and political reform in this country.

We ended up calling it the Cummings Creek Compact, the agreement we came up with that weekend to continue the work we'd started with the campaign. We talked about forming a new group and starting a blog—Change for America—to help us organize people, raise money, and advance the cause of empowering the American people again. Change for America would be an online clearinghouse for groups committed to expanding democracy, a place for issues to be raised, agendas to be written, constituencies to coalesce. As we wrote in the compact:

> Change for America will be a national organization that unites progressive communities and sets an agenda of meaningful reform. The values that shaped our campaign are the same values that formed the moral foundation of our American democracy—and we carry those values today. We are committed far beyond a single election.
>
> Our immediate path is clear: we must defeat George W. Bush and elect a new president, we must infuse elections at every level with the same commitment that built this movement—and you must decide how to do it.

Something that Thomas Jefferson said seemed particularly apt to me that day: *A little rebellion is a good thing*.

God forbid we should ever be twenty years without such a rebellion. . . . What country can preserve its liberties, if its rulers are not warned from time to time, that this people preserve the spirit of resistance?[1]

By my count, it has been forty years since the heyday of the civil rights movement, the last real democratic rebellion in this country. We're overdue.

Just before the sun went down that Sunday, I put the Cummings Creek Compact to a vote. If we do this thing, I asked, how many of you would be willing to continue the fight? I thought for a moment about my daughter's class. But when I looked around the room this time, everyone's hand was raised high in the air.

HAVING FAITH IN STRANGERS AGAIN

During the campaign, Neil Ambercrombie, the Hawaii congressman, came up to me after attending a Meetup in New York and told me that it had dawned on him how, for all the futuristic buzz around the Dean campaign, what we were actually up to was as old as the country itself. "I don't know if you realize what you're doing," he said, "but the community you've built is really about having faith in strangers again."

A single political campaign is only a small scratch on the surface of what this technology will eventually accomplish. We'll see the real power and value of this return to civic life when someone like Howard Dean is elected because of all those strangers coming together. When the next candidate begins *governing* using the Internet—that's when it will get interesting.

Imagine the presidential candidate who is able to continue Dean for America's exponential Internet growth . . . when it grows to two, then four, then six million online Americans—which Dean would have had if he'd been the nominee—all linked up on the Internet. At that point, the election will only be the beginning. That president's mandate would be a living, breathing thing.

---

[1] Jefferson's letter to William Smith, 1787.

Polls show that a majority of Americans want government to provide some level of health care, especially for children. For decades, politicians have talked about this. So why don't we have it? Because the top five pharmaceutical companies gave $11 million to Republican candidates last year. Because there are more lobbyists for the drug companies and HMOs in Washington than there are congressmen. Because whenever someone even brings up health care as an issue, the special interests spend tens of millions of dollars on manipulative advertising campaigns to scare people into thinking they won't get to choose their doctors. Remember Harry and Louise?

But that's *before* the president of the United States shows up in Washington with the e-mail addresses of six million of his closest supporters. Before the president vows to govern the way he'd won—by tapping into the will of the American people. Before he drops them all a note that says, "Hey, if you really want health care, I need your help. Go to your computer right now and e-mail your congressman and tell him that you don't want him listening to the pharmaceutical lobby, that you don't want him listening to the HMOs. Tell him that you want him to listen to *you*. And here are the names of those drug companies who don't want us to provide affordable health care. And here is a list of the products they want you to buy and the generic alternatives to those drugs."

Meanwhile, the United States is racking up the largest spending deficit in its history. As politicians promise to lower our taxes, provide health care, a strong military, early childhood education, and a hydrogen car in every garage, our country is borrowing itself into a deep hole, passing the cost of this on to our children. Your kid's share of the deficit is something like $180,000 now. You would never do that in your own life. You live within a budget. You'd never knowingly buy a $5 million mansion on a ninety-year mortgage with a balloon payment that is passed on to your kids.

So why are we letting our government get away with this? Liberals don't like it. Conservatives don't like it. Who *does* like this?

Lobbyists. But their day is coming to an end.

Issues like health care and the deficit are naturals for government to solve using the Internet. No special interest will be as powerful as six million empowered Americans. As soon as a candidate uses the Internet to declaw the lobbyists and make our leaders be responsible by galvanizing the American people into a force that can be heard . . . it's over. The whole thing is over. We'll have won.

This is closer than you think. I believe the 2008 election will be the first national contest waged and won primarily over the Internet. We are, right now, in the midst of sweeping aside shallow, cynical broadcast politics for the politics of ideas and issues that will define the Internet age.

So what will the new landscape look like?

At some point, of course, there will be convergence. One box, One screen. You'll check your e-mail and order your groceries and check your child's homework all on the same screen. That might be the most dangerous time for this burgeoning democratic movement—the moment when the corporations and advertisers will threaten to co-opt and erode the democratic online ethic. The future may well hinge on whether the box is dominated more by the old broadcast rules or by the populist power of the Internet.

Thankfully, that decision won't be made by the courts or by the Congress, by the president or by the corporations. We will make that decision. I've talked a lot in this book about what the politicians have to do, what the corporations have to do, what the institutions of our society have to do to survive. But, in the end, this isn't their technology.

It's ours. From this point on, it's up to us.

The next decade is probably inevitable. First movers on the Internet will continue to have a huge advantage. The companies that dominate will be the ones that manage to build online communities, Joi Ito's parks and gardens—a world of new eBays and Googles. Look at all the things you do on the Internet today that you didn't do ten years ago—buying airline tickets and books, sending e-mail to friends, reading news and entertainment sites, downloading music—and realize that in ten years that list will grow virally, exponentially. Look at the things that were ahead of their time just a few years ago—smart mobs, online groceries, and yes, Howard Dean—and know that they will be the norm in ten years.

Weinberger, in *Small Pieces,* theorizes that democratic forces are always shaped by their opposite, by the things they rise up in response to. That democratic forces rush in to fill the space that's needed. American democracy originally rose up in response to British aristocracy and became its opposite. This is why I am convinced that Internet politics and government will be defined by its opposite, broadcast politics, and by its potential to fix many of the problems broadcast politics creates:

- Civic disengagement. The Net builds communities and brings people together, providing the first reversal of trends reported in Robert Putnam's alarming book *Bowling Alone*—the isolation of Americans, the death of participatory politics, and the unraveling of the fabric of critical social and civic structures.
- The dumbing down of the American electorate. We're all tired of those studies that show half of schoolchildren can't name the president or the capital of their state. For the past fifty years, people have assumed they have to compete with the shallow flash of TV to get anyone's attention, but the Internet is growing exponentially with a very old-fashioned recipe: reading and writing.
- The insidious corruption of our politics and our government due to the disproportionate influence of wealthy donors, special interests, and corporations. The Internet shines a light on these dark recesses and quickly organizes millions of Americans *cheaply,* without relying on billionaires who want something for their money. Unlike TV ads, which can cost millions, on the Internet, all you need is a web site and working fingers.
- Various other diseases of broadcast politics, including attack ads, governing by sound bites, and celebrity politics. When the Internet has become the dominant information media in this country—in the next few years—TV will go back to doing what it does best, entertaining us. TV is great for *Law and Order*. It's not so good for making laws and keeping order.

I suppose to some, my vision of the Net's impact on government and politics makes me sound like a hopeless optimist, one of those dewy-eyed computer Utopians who went around in the 1990s imagining a future of virtual vacations and online grocery shopping—or those 1960's futurists imagining jetpacks and flying cars.

Fine. I plead guilty. But it doesn't mean I'm wrong.

I believe in democracy above everything. I believe that the best government is one in which the people are allowed to govern themselves. I believe, as Thomas Jefferson did, that Americans must participate fully in democracy for it to work. And I believe the Internet is the best tool we've ever created to help achieve that.

This generation of activists is being defined by what they accomplish using the Internet, just as surely as my generation of politicians and strategists was defined by and, eventually chained to, the television. But while TV was a medium that rendered us dumb, disengaged, and disconnected, the Internet makes us smarter, more involved, and better informed. The Internet was designed to foster cooperation; it's built on a foundation of shared innovation.

In two years, some kid will look back at the way we used the Internet on the Dean for America campaign and be shocked at how primitive it was; Meetups and blogs will look as quaint to him as the John Kennedy dancing political button commercial looks to me now.

This is the beginning of *their* time. When they use this technology to transform our country in a thunder roll of democracy and change, nothing will ever be the same.

The world ahead won't be perfect. There will be new problems. Crooks, frauds, and sex offenders can find their way online just as they can come into your neighborhood. Viruses, invasion of privacy, and copyright violations—there are a million pitfalls. Rumors will fly. Misinformation will sneak into the system and some people will blame the medium, the same way they blamed radio for the panic that resulted from Orson Welles' *War of the Worlds* broadcast about a massive Martian invasion.

But I'm not worried. The Internet itself is the best solution to the problems it will create. It is a problem-solving medium that runs on the combined power of a billion computers, on the remarkably innovative power of people.

The naysayers had their decade of smug denial. Now, the world has changed. Now, the proof of America's transformation is everywhere you turn. Now, there are only two paths: Recognize this new technology as the redefining structure of our lives . . . or continue to believe that it's only another gadget. Join the ranks . . . or hunker down behind the gates. And in the end, it doesn't really matter what you do.

The revolution began yesterday.

THE BLOGOSPHERE AWAKENS

On December 5, 2002, a group of Strom Thurmond's friends, family, and Republican members of Congress gathered for the hundredth birthday party of the oldest and longest-serving member of Congress. Various people gave speeches, and then Mississippi Senator Trent Lott, the Senate majority leader, rose to speak: "I want to say this about my state," Lott said, "When Strom Thurmond ran for president, we voted for him. We're proud of it. And if the rest of the country had followed our lead, we wouldn't have had all these problems over the years either."

Witnesses said that Lott's remarks were met with stunned silence and audible gasps, as party guests came to the logical and uncomfortable conclusion that Lott clearly must have known that when Thurmond ran for president, he had only one real issue: preserving segregation. Most of the major newspapers and broadcast news outlets missed this logical conclusion, though. They ran small stories about the event, but only one, *ABC News,* mentioned Lott's outrageous comment. The rest did features on the old boy's birthday party. One reporter said reporting Lott's statements didn't seem to fit the tone of the birthday story. Five years ago, that would have been the end of the story.

But in 2002, the blogosphere was there to pick up the ball right where the media dropped it. When Lott's quotes appeared on an ABC News web site,[2] two of the Net's most influential blogs, Atrios and Talking Points Memo (Joshua Micah Marshall's site), jumped all over the story. Soon other blogs were talking about it and the controversy spilled over to more traditional web sites like Slate, until the old media finally had to take notice, and chipped in almost a week later with a 1980 speech in which Lott said essentially the same thing. By December 20, Lott had resigned his leadership position—the first time in history a Senate Majority Leader had ever been forced out. All because of a handful of bloggers.[3] As one writer put it, the Internet had "its first scalp."

A few months before the Dean for American campaign got rolling, the Lott story sent a shockwave across the country for people who hadn't realized just how powerful the Net has become in politics, and in framing the events of the day. After all, there were only a couple of million bloggers in the whole country and these were not people with access to the institutions

---

[2] Kudos to ABC's Marc Ambinder for his reporting. Marc would later be the ABC News "embed" with the Dean campaign.

[3] For a detailed study of the effects of blogging on Lott's resignation, see "Big Media" Meets the "Bloggers," written by Esther Scott for Alex Jones, director of the Joan Shorenstein Center on the Press, Politics and Public Policy, Kennedy School of Government, Harvard University. http://www.ksg.harvard.edu/presspol/Research_Publications/Case_Studies/1731_0.pdf.

that have traditionally driven public opinion in the United States—newspapers, magazines, and television networks. And yet, with blogging still in its infancy, the top four bloggers had the combined reach of the *New York Times*. Dan Gillmor describes this new class of media in his recent book, *We, the Media: Grassroots Journalism by the People, for the People:*

> Outsiders of all kinds can probe more deeply into newsmakers' business and affairs. They can disseminate what they learn more widely and more quickly. And it's never been easier to organize like-minded people to support, or denounce, a person or cause . . . modern communications have become history's greatest soapbox, gossip factory, and, in a very real sense, spreaders of genuine news.

The little-known secret in newsrooms across the United States is that *right now* reporters are beginning every day by reading the blogs. They're looking for the pulse of the people, for political fallout, for stories they might have missed. The terrorist attacks of September 11, 2001, were a key moment for the growth of the Internet as a source of news and information. A Pew Research Center Internet study found that in the immediate days after September 11, just 3 percent of Americans used the Internet as their *primary* source of news and information. Less than two years later, as the United States was preparing for war with Iraq, that number was 26 percent. From 3 percent to more than a fourth of the population in *two years*. And even if it's not their primary source of news, 77 percent of Americans now say they use the Internet to interact with the news of the war in Iraq. The key word here is *interact*. These people aren't just reading web sites for news of the war. They're e-mailing stories to one another, posting messages, blogging, and writing electronic letters to the editors of publications.

Like television's ascension fifty years ago, the Internet's impact is not just in the way people get their information. It's affecting what they do with that information and that, in turn, is altering the way our leaders behave. Television certainly gave us visceral images—Kennedy's assassination, the moon landing, Vietnam—but it lulled us to sleep as a culture. The Internet, a medium of reading and writing, is waking us up.

It's also having a profound impact on the events themselves. As television changed the way events were covered, the events themselves began to change (the Vietnam War and Watergate, for instance). Similarly, as it expands, the Internet is creating space that never existed before—like an electronic Big Bang.

The War in Iraq will go down as a watershed moment in the ascension of the Internet. During the initial invasion, with reporters embedded with American troops, the broadcast coverage resembled cheerleading and a large number of Americans lost faith in it. When George W. Bush was making a case for Iraq's weapons of mass destruction, where were *Nightline* and *60 Minutes*? Now that it's becoming clear Bush *always* wanted to go to war, the media's football-game coverage of Iraq (while any substantive discussion of the underlying issues was ignored) and the Congress' rubber-stamp approval of it seem almost profane.

The media's role was not much different than the military-censored war coverage of World War II. The difference was that skeptical Americans, hungry for real debate, had somewhere else to turn. They could go on the Internet and read other newspapers, or listen to the BBC, or read blogs from people in other countries. And the more homogenous the reporting became about the war, the more it drove people to the Internet, helping to grow the monster that is going to kill it.

Newspapers are declining. Television viewership is declining. Where are those people going? They're turning on their computers, in ever increasing numbers. Newspapers and television news programs are fighting over a diminishing group, squeezed ever further and further into the center while they lose millions on either end to the Internet.

And so, at a time when politicians assumed that they couldn't criticize the war and the media was afraid its ratings would fall if it asked too many questions, real Americans were doing both of those things online. They wanted choice. And that was one thing that Howard Dean gave them when no one else dared.

Just a year later, we're already taking for granted the myriad ways in which the Internet and telecommunications are changing the information we receive and altering the political landscape. When the first photos of flag-draped coffins of American soldiers killed in Iraq were published in the United States, few people knew that it was an Internet web site that first sued for, and eventually won, the right to publish those pictures.

Amid the outrage over the photographs of American abuse of Iraqi detainees at Abu Ghraib prison, few stopped to ponder the other significance of those incendiary photographs. These were not official military photos, or photos taken by journalists—the only photos that came out of earlier wars. These were pictures taken on digital cameras by American soldiers, like thousands of others, some of which were sent over the Internet to friends and family back home. As the British Broadcasting Service reported: "The Internet has been acting as an unofficial clearing-house for all sorts of unapproved images of conflict in Iraq."

Now, because of the wide dissemination of these technologies, Americans are seeing the truth behind our own propaganda—the evidence of abuse by Abu Ghraib prison guards—before some of the highest-ranking people in the government. In his testimony before Congress, Defense Secretary Donald Rumsfeld angrily admitted there was nothing he could do about the flood of unrestricted information and photos.

"We're functioning with peacetime constraints, with legal requirements, in a wartime situation in the Information Age," Rumsfeld said, "where people are running around with digital cameras and taking these unbelievable photographs and then passing them off, against the law, to the media, to our surprise."[4]

It's almost as if Rummy fell asleep in 1970 and woke up in a new world. He's not the only one.

The new paradigms of fluid media and shifting political power do not stop at the U.S. border, either. In some places, these new technologies are even more influential than in the United States. In China, people have used the Internet and digital telephones to circumvent their government's and media's denial of the SARS and AIDS epidemics, finally forcing the government to confront the deaths of its people and to do something about it. In the Phillipines, the People Power II movement (one of Howard Rheingold's "Smart Mobs") used cell phone text messaging to organize nimble and powerful demonstrations that led to the ouster of President Estrada in 2001.

In May 2004, I was a principal speaker at the Internet Global Congress in Barcelona, Spain. I met my counterparts from all over the world, like Veni Markovski, of the Internet Society in Bulgaria, who had helped organize people in that country for a decade. In 1995, he said, there were only about fifty people online in Bulgaria. A few years later there were 2,000. A

---

[4] Robert Plummer, "U.S. Powerless to Stop Iraq Net Images," BBC News Online (May 8, 2004).

few years later, it was a million. Then 1.6 million. And because his group was there organizing these people into a community from the beginning, they have become a powerful political constituency that no candidate or issue can afford to ignore. Now, they are 20 percent of the Bulgarian population and growing fast.

One day, Veni told me, he met with a former KGB agent in Bulgaria (although he swears there is no such thing as a *former* KGB agent). "I am happy to see you talk so much about changing things using the Internet," the former KGB guy said. "What worries me is that you actually intend to change things."

Yes, we do.

In America, for two hundred years, we have relied on some version of the media to interpret the events of the world for us, and at the same time to explain our governments' role in them. The past fifty years have been dominated by one medium—television, as we sat around, waiting for a square-jawed talking head to tell us what to think, in between commercials for dandruff shampoo.

People are no longer waiting for the media or the government to *give* information. Now they are going online and getting it, and then disseminating it. And with that information, they are gaining power.

And that . . . is where you come in.

A warning for hardcore Republicans: This might be a good place to stop reading. Get a beer. Go for a walk.

I really don't mean to indulge in partisanship here, but come on. I'm forty-eight years old and I've been a Democrat my whole life. As I sit on my Maryland farm in the spring of 2004, there is still some unfinished business out there. See, I genuinely believe that the administration of George W. Bush is the embodiment of everything our founding fathers feared most—a government that has finally capitulated to the corruption of money, a government that exists primarily to enrich its financial backers, that blatantly lies about public policy, and invents reasons for going to war.

Just to be clear, I don't necessarily blame George W. Bush. He didn't start this era of decay of government. He's just the last in a long line of people who have chipped away at it for generations until there's nothing left.

So don't think of what's about to come as partisan politics.

Think of it as . . . Jefferson's revenge.

YOU HAVE THE POWER

We are at a critical moment in American history.

We are at the point that our founding fathers feared the most—when economic power seizes political power in this country.

America's poet laureate of liberty, Thomas Jefferson, spoke often of this danger. Jefferson wrote that the corrosive nature of money in politics virtually guaranteed that one day, "corruption . . . will have seized the heads of government, and be the voices of the people, and make them pay the price."[5] In fact, Jefferson might have been describing American politics in the year 2004:

> Unless the mass retains sufficient control over those entrusted with the powers of their government, these will be perverted to their own oppression, and to the perpetuation of wealth and power in the individuals and their families selected for the trust.[6]

That's why I believe this very moment in time is exactly what Jefferson feared most—a perversion of government built for its own "perpetuation of wealth and power."

Don't get me wrong. I love capitalism. So did the authors of our founding documents. They knew that capitalism was a wonderful thing, the engine of this country's growth. But they also knew that democracy must govern capitalism, not the other way around. They knew that our economic system had to be subservient to our government. Left unchecked, they knew that capitalism quickly devours everything in its path, including democracy.

This is exactly what has happened to America over the past quarter century.

We now live in an age when public offices go to the highest bidder, when the Bush administration routinely and brashly trades government

---

[5] Thomas Jefferson, *Notes on Virginia*, 1782.

[6] Jefferson's letter to M. van der Kemp, 1812.

contracts, appointments, and access for political donations. We live in an age when the executive branch behaves like a $300 million shadow corporation, its board of directors is made up of the wealthiest one quarter of 1 percent of Americans, whose only aim is the perpetuation of their own wealth and power. When pharmaceutical lobbyists are allowed to write drug policy and oil lobbyists write energy policy. We live in a time when our government gives these special-interest-authored bills Orwellian names like the Clear Skies Act (which naturally allows more pollution), and the Healthy Forest Act (more clearcutting).

We live in an age when political discourse is reduced to short, cynical TV commercials, when Americans feel increasingly alienated from their own system of government. When more than half of eligible voters don't bother going to the polls, 70 percent of Americans are unhappy with *both* political parties, and 60 percent of Americans routinely assume that every candidate is lying to them. Perhaps most distressing of all, we live in an age when two-thirds of young people tell pollsters that they are powerless to change anything . . . except the channel.

The Founding Fathers anticipated that dark days of democracy would inevitably arrive, but they also understood there was an inherent brilliance to our system that would allow us to find our own way out.

Jefferson got it:

Whenever the people are well-informed, they can be trusted with their own government. Whenever things get so far wrong as to attract their notice, they may be relied on to set them to rights.[7]

James Madison got it, too:

Knowledge will forever govern ignorance, and a people who mean to be their own governors must arm themselves with the power which knowledge gives.

The Age of the Internet could not have arrived at a better moment.

I don't care what political party you belong to—if any—you have to admit, there is something wrong when six hundred thirty-one fundraising "pioneers" and "rangers"—billionaires and multimillionaires—can

---

[7] Jefferson's letter to Richard Price, 1789.

bundle together obscene sums of money and use it to buy our government and perpetuate their own wealth and power—while our nation's problems are ignored.

Our Founders feared this would happen—and that is why they placed the power to elect and change our government in the hands of the people. We've had the *right* to fix this broken, corroded, and corrupt system since the earliest days of our Republic. Two hundred and eight years later we finally have the *tools* to do it—the tools to bring us together in common purpose to reclaim what is rightfully ours, beginning with the White House, the Senate, and the House of Representatives. We have it in our power to reclaim our government and with it our national purpose.

It would be easy to do. If six hundred thirty-one wealthy individuals can raise more than a hundred million to elect George Bush—it will only take two million Americans contributing less than a hundred dollars each to defeat him.

Call it Jefferson's Revenge.

Call it the $100 revolution. It doesn't matter what you call it, now is the time for it to happen.

No one can change America for you. You have to do it. If just two million of us give what we can in time and money, we can take back the White House and the Congress. A hundred bucks from two million Americans. That's all it will take to send George W. Bush packing. To return America to the principles that it was founded on.

And the great part is this: after you raise the $200 million, I don't even think you'd need to spend it. Just the act of raising two hundred million dollars from two million Americans over the Internet will be enough to turn the country on its head. The mobilization of that many Americans . . . the sheer engagement of that many people in politics again will signal to lobbyists that their days of writing policy are over, that the days of a few hundred billionaires and millionaires choosing the president are now done.

In the beginning of this book, I wrote that we had misnamed this era *The Information Age.* I said that it should more appropriately be called *The Empowerment Age.* This is what I meant: The Internet is the most democratizing innovation we've ever seen—more so than even the printing press. There has never been a technology this fast, this expansive, with the ability to connect this many people from around the world. If Madison was right, and the people can only govern when they can "arm themselves with the power which knowledge gives," then the Internet is the first technology that truly gives people full access to that knowledge—and empowers them with the ability to do something with it.

We find ourselves, in the early morning of the twenty-first century, at one of those moments that Howard Dean spoke so eloquently about in Iowa: a time in this country when it's my responsibility as an American citizen and your responsibility as an American citizen to work together, collectively, cooperatively, for the good of us all.

We can do this now because we are more connected than at any time in our history. But this thing that connects us is more than a mesh of wires and optic links, more than a world of web sites and blogs and e-mail addresses. We are connected by our birthright as Americans and by the very fiber of democracy.

Together, I believe we can accomplish anything, if we just keep one idea in mind. One principle. Four simple words that still echo across America:

You Have the Power.

AFTERWORD

ere I was again. After working for Howard Dean in 2004, I had
vowed never to work in another presidential campaign. And for the
first frickin' time in my life I meant it. So why was I breaking that
promise—to myself more than anyone else—and getting into the heat of
the 2008 campaign?

Who was I kidding? This is who I am.

The Dean campaign had been one of the great experiences of my
life. Afterward, people would stop me in airports and thank me for
helping to strengthen our democracy. And I always responded the same
way: "No, thank *you* for getting involved and making a difference."

Amazing things happened to me because of my work in that campaign—things I never could have dreamed of. I was invited to Moscow
to work with organizers of political parties and others on how to use
the Internet and cell phones to connect with each other and start giving
Russian citizens a stronger voice. From the Persian Gulf state of Qatar
to Nigeria in Africa, from working for Prime Minister Tony Blair in
the United Kingdom to helping guide Romano Prodi's successful bid
to become prime minister of Italy, I knew I was there because of my
work on Dean for America. This boy who grew up so poor he had
water in his cereal instead of milk, who never graduated from college,
was now a Fellow at Harvard's Kennedy School. It had to be a mistake!
Even corporate America wanted to know what I was thinking—and
they didn't get pissed at me when I told them, the way so many in my
own party did.

The Dean campaign had been good for me. I was proud of my
work, but it wouldn't have been possible if Howard Dean hadn't given
me the chance, and if hundreds of thousands hadn't gotten involved in
a presidential campaign that pioneered a new way of doing things. We

proved that ordinary citizens had a voice again. We may have lost—but nothing would ever be the same.

I laugh at the assholes who carp—from the safety of their blogs or their precious seats on some cable talkfest—about the win-loss records of presidential campaign operatives. The chattering heads who never laid everything they had on the line for two years or more, only to have their heads bashed against the rocks of the most unforgiving process ever devised—a campaign for president of the United States.

Don't you get it? I want to tell them. *Almost no one ever wins that game.*

You fight to take your party and your country in the direction your candidate believes we must go. And that's if you're lucky enough to have a candidate who wants you at the helm and you're stupid enough to want the job. Did it ever dawn on anyone that Bob Shrum may have been in so many presidential campaigns because he was that damn good? Guess what? He was that damn good.

But no matter how good you are, how good the candidate or the campaign is, a presidential campaign can all go south in the time it takes to make that one six-second sound bite that will last forever. You never sleep, you die a thousand deaths, you hear yourself gaffing and pray you didn't just take the whole damn thing down. You don't take care of yourself—because if that's what you want to do, you don't need a doctor to tell you not to join a presidential campaign. And the pressure? It's like diving to a depth of three hundred feet without a wet suit—and staying there for the whole campaign.

I knew all that in 2003 when I joined the Dean campaign. I knew it in 2007 when I sat down to meet with John and Elizabeth Edwards. I knew how tough the odds were. But then again, changing the odds had always been what inspired me.

As a teenager, I had dreamed of being Gene Pokorny, the grassroots gunslinger I'd read about in Hunter Thompson's *Fear and Loathing on the Campaign Trail.* Then one day I opened up *PC World* to a piece with the headline "Meet the Next Generation's Whiz Kids," and read about a young blogger named Stephen Yellin who was being described as "the Trippi of the future." They were the sweetest and, at the same time, the scariest words I'd ever read. *Wait!* I thought. *I'm not dead yet!*

Don't get me wrong—it meant a lot to me to think that I'd inspired young people in any way to get involved in making a difference. Robert

Kennedy, Gene Pokorny, and Hunter Thompson had all conspired, wittingly or not, to trigger that fateful decision in me. But it's not quite as romantic as Hunter made it sound—and whatever good my thirty years of working for my party and my country may have fostered, it came at a terrible cost in terms of my health.

When my doctor first diagnosed my type 2 diabetes in 1990, he urged me to get out of politics altogether. It was clear that my life on the campaign trail had contributed to the onset of the disease. But how could I stop doing what I loved? So I was more careful, and through the 1990s I pretty much avoided presidential politics. It was a huge battle, but somehow I was able to work in state and local campaigns, consult for high-tech companies, and keep the damn disease in check.

Until the Dean campaign. Those fourteen months did me in. Endless weeks spent eating junk food on the run, on two hours of sleep a night, under incredible stress, was a prescription for explosive diabetic complications.

A year after the Dean campaign was over, I was running four to six miles a day, four or five times a week. I was kicking diabetic ass, dropping the pounds, and trying to get back in pre-Dean shape.

Then, suddenly, I hit the wall. I could not take another step without searing pain racking my body. I crawled home from a run one day, went back to the doctor, and discovered that I had developed diabetic neuropathy—the inevitable result of a year or more of self-abuse in the name of a cause. Those crazy months on the Dean campaign would now be with me for the rest of my life, in a way I wouldn't wish on my worst enemy. I'll skip all the medical happy talk—you can Google it—but what happens with diabetic neuropathy is that your nerves basically go haywire in all kinds of mysterious ways. Mine decided that the only signal they knew how to send was pain.

Mike Ford, a man I've revered and learned from throughout my political career, had developed diabetic neuropathy long before me. When I asked him how he dealt with it, Ford looked at me with those knowing Irish eyes of his and said, "Bro, it's amazing how much pain you can get used to. After a while it's like it isn't there." That was a comfort. We'd been brothers for years, the least he could do was lie to me.

I hesitate to say this here, but in the first two months after I was diagnosed—two months of frickin' bees stinging me all day long, of dagger pains and glass shards sticking me so relentlessly every night that I couldn't sleep—I entertained my share of sniveling thoughts about ending it all. But I have a beautiful and amazing wife, and three wonderful kids, and I could never do that to them. Not to mention this heroic image I walk around forcing myself to uphold, which allows me to soldier on—that and an amazing array of antiseizure medicines. No one, including the pharmaceutical companies that make them, knows exactly why or how those things work—but they do. So now I can drive my wife crazy for another thirty years or so.

It turned out that Mike Ford was right—it is amazing how much pain you can get used to. Over time I've learned to mask the momentary flare-ups that can still cut through the medicine, so that no one else in the room can tell that thousands of invisible bees have just decided to use me for target practice.

No one, that is, except for Elizabeth Edwards.

In that February of 2007, I had agreed to spend a few hours with John and Elizabeth Edwards, to offer them my advice on how they should approach the 2008 presidential campaign. I'd been talking with them for about an hour in their home in Chapel Hill, North Carolina, when the bees started doing their thing—stinging the bejesus out of me. I masked it as usual, answering their questions without a wince or a twitch—or so I thought—when Elizabeth stopped the conversation.

"Joe, do you need to take some medicine? Are you in pain?"

Wow, I thought. *She's good.*

I explained that I had something called neuropathy. It flared up from time to time, but I'd be okay—no worries.

That's when Elizabeth told me she had neuropathy too. "It makes my hands and feet numb, so I can't feel anything," she said. "I hate it."

By now, the bees were swarming every inch of me. I became so focused on my own pain that I lost sight of who I was talking with, and what she was going through.

"I wish I had the kind you have!" I blurted out.

John Edwards got up from the table. I was sure I'd said something offensive, but I had no idea what it was. By this time, the bees had turned into chainsaws, and I kept digging myself in deeper. But Elizabeth didn't miss a beat. She asked me about my pain, about how tough it must be for me—and her concern was so genuine that I suddenly realized how crazy it would be for me to take on another presidential campaign in this condition.

As soon as it was politely possible, I thanked them both and made for the door. I was impressed with their commitment, and I told them so. "I really think you want to change things," I told them. "But I've got a wife, three kids, some dogs, some cats, and a one-horned goat, and I can't possibly work on another presidential campaign."

I gave Elizabeth a big hug good-bye and made it to my rental car so fast that I was driving away, pedal to the floor, before I realized I hadn't said good-bye to Senator Edwards. I still remember looking into my rearview mirror and seeing a bewildered John Edwards waving me good-bye from his front doorstep.

I, on the other hand, was ecstatic. I had slipped the bonds of presidential politics and managed to dodge my shot at working a 2008 campaign. I was sane after all!

I was doing mental somersaults down the aisle of my plane home . . . when it hit me. Elizabeth Edwards must have contracted her neuropathy from her battle with breast cancer. *What an insensitive idiot I am.* I beat myself up for the rest of the flight.

I may be dumb, I finally decided, but at least I'm smart enough to avoid taking on another presidential campaign.

Or was I?

I knew what was coming—it was that same feeling I'd had back in 2003, when I knew the time was ripe for the Internet and politics to come together and create a new kind of presidential campaign. Now, in the early days of 2007, I could feel it in my bones: we were close to the tipping point. Some people may not have liked Howard Dean or his campaign. Some people might have been slow on the uptake. But I knew this: If Dean for America had failed to wake you up to the potential for bottom-up transformational change in our country in 2004, by the end of 2008 you'd be wide-eyed with amazement and wouldn't be able to sleep.

The tools we used in the Dean campaign through most of 2003 were downright primitive compared to what campaigns would have to work with in 2008. In the Dean days we had Meetup.com to organize meetings once a month. We built our own "get local" tools, which allowed people to organize themselves by zip code. We had a few key blogs to spread the word, and our webmaster, Nicco Mele, had that wicked collection of yo-yos. That was pretty much it—that and a bunch of daring pioneers willing to push the envelope, no matter who laughed.

By the time the 2008 campaign was heaving into view, though, no one was laughing anymore.

Every day since the 2004 campaign, those primitive tools, techniques, and strategies we'd developed were spreading. State and local campaigns, national parties, nonprofits, and corporate America had begun using them, developing them, and improving them. And that was just in the good old U.S. of A.

In the 2006 midterm elections, the network of progressives on the Internet had proven smarter and in many ways more powerful than the official party committees in Washington. In Montana, the Democratic Senate Campaign Committee had virtually handpicked the state auditor, John Morrison, as *the* candidate for U.S. Senate. But the party's netroots had placed early bets on a little-known state senator named Jon Tester. After Stephanie Schriock, the miracle-working finance director of Dean for America, took the helm of the Tester campaign, she managed what became a brilliant upset for Montana's Senate seat.

A growing and active online community helped create early buzz for Tester, generating online contributions as he gained traction in Montana, won his primary, and proved the Washington crowd wrong. When video of his general election opponent, Senator Conrad Burns, surfaced showing the incumbent nodding off to sleep during a Senate hearing, hundreds of thousands viewed the untimely nap on YouTube—bringing more attention to the race and even more campaign contributions to Tester.

But people also figured out that these new tools couldn't perform miracles—that none of it mattered if you didn't have the right candidate and a disciplined strategy. Jon Tester was the personification of the Montana values Conrad Burns had lost touch with. And Schriock proved to be one of the best strategists and managers in the Democratic party—she became the chief of staff to a brand-new U.S. senator and suddenly I had to be careful what I called her . . . the LBSOS!

Two new Internet companies, Blue State Digital and EchoDitto, were launched by former members of the Dean campaign's web team. I had laughed with them, brainstormed with them, yelled at them and with them. But now they really were two of the best teams in baseball. They were leveraging everything we had learned and built together, and now they were armed with a bunch of new stuff—social networking and viral media tools—that we could only have dreamed of in the Dean days.

By 2006, I realized that the story of the Dean campaign had turned the old adage "Victory knows a thousand fathers, but defeat is an orphan" on its head. A new industry had sprung up in the political world, with firm after firm touting its knowledge of politics and the Internet—and a shocking number of them claiming to have been there in Burlington working for the Dean campaign. We may have lost the race, but people were making a shitload of money saying they were us, or that they'd been the ones who told us about this thing called the Internet in the first place.

Still, everywhere you looked, there were new tools coming online that were making it easier for Americans to connect and communicate with each other. Social and viral media sites like YouTube, social networking sites like Facebook, MySpace, and Bebo, and mobile networks like Twitter had been all but nonexistent in 2004. Before 2008 even began, most were household names.

In some cases, what we considered "really cool ideas" turned out to be mini-revolutions of their own. We'd developed Dean TV, a 24/7 broadband video wonderland where anyone could post their own user-generated video showcasing their favorite issue or why they had joined the campaign. No one got rich off of it, but we'd created our own You-Tube before YouTube created YouTube. People used it to create their own music videos, mashing up speeches and music at a time when the press still thought mashing was something you did with potatoes.

The campaigns of 2008 would have social networking sites and viral video platforms built by entrepreneurs and funded by venture capitalists, scaling and growing at an incredible rate as more and more Americans discovered the genius of each other and the impact a single citizen could have with just a video cell phone. Where the hell were those things in 2004?

It was amazing to see who was using these tools. We had gone from about 1.4 million blogs by the end of the Dean campaign, to more than 77 million by early 2008. People were posting their thoughts, their pictures, their ideas, and their take on the news like never before, and millions more were commenting on those posts. If Howard Dean had generated commentary, buzz, and actual commitments from participants in a blogosphere of 1.4 million blogs, there was no telling what could happen with 77 million.

But blogs weren't just growing in number. They were changing journalism—changing who wrote the news, what became news, and how corporate America viewed communications.

In the waning days of the 2004 general election campaign, a poster with the screen name "Buckhead" had posted on the conservative site FreeRepublic.com that documents used in a report by Dan Rather on CBS News were forgeries. The documents allegedly showed that George W. Bush disobeyed orders while in the National Guard, and had undue influence exerted on his behalf to cover up his shabby record. Conservative blogs with names like Little Green Footballs and Powerline spread Buckhead's credible posts that the documents were false. It should have been a signal to everyone: Buckhead was right and Dan Rather was wrong—so disastrously wrong that he was soon forced to step down as the anchor of *CBS Evening News*.

The game had radically changed, and politicians, journalists, and corporations alike were scrambling to catch up. It turns out that everyone's an expert at something—at plumbing, or computer fonts, or 1970s typewriter technology. Suddenly, though, those experts had blogs, and those blogs were read by hundreds of thousands, sometimes millions of people. The world was full of people who might know more about the specifics of any given story than anchors or politicians or CEOs—and who weren't afraid to call you out if you got it wrong. Buckhead, as it happened, was an expert on IBM Selectrics and type fonts, and Dan Rather wasn't. And all at once we realized: someone with a blog and some knowledge, or a cell phone camera in the right place at the right time, could change the course of an election, or challenge a corporation's image.

Even after Buckhead, though, there were more surprises in store. Who would have foreseen that one of journalism's major awards would be won by a blogger?

In 2007, Joshua Micah Marshall of the Talking Points Memo blog won the George Polk Award for Legal Reporting for his coverage of the U.S. Attorney firing scandal. The Bush administration was dismissing U.S. Attorneys for political reasons and Marshall blogged away about it—uncovering the story, reporting it, and driving it when no one in the "mainstream media" would touch it. Some of the mainstream press even ridiculed Marshall for pursuing a story they said wasn't there. Now the Polk Award committee was praising Marshall for his "tenacious investigative reporting," which had "sparked interest by the traditional news media and led to the resignation of Attorney General Alberto Gonzales." Apparently there were those in the Bush administration who thought they could get away with murder—and they would have, if not for a blogger.

Frickin' amazing.

In 2006, IBM began encouraging its more than three hundred thousand employees to start their own blogs. Big Blue *got it:* if each blog had an average readership of just ten people it would mean that IBM's employees would be communicating with more than three million people a day. Big Blue wasn't going to get caught off guard by the changing communications environment, and they'd also learned something we learned in the Dean campaign—that listening, really *listening* to people and their ideas, works. Just like we screwed up by failing to post a "Puerto Rico for Dean" sign—and then fixed our mistake after reading about it on our blog—IBM would be able to listen for early warnings about mistakes, or pick up a good idea and act on it, or simply do some good as part of the growing online community. At a time when many corporations were running away from the blogosphere, IBM was running toward it.

And while the blogosphere was exploding, the Dean Web team alumni weren't exactly standing still. Rick Klau, one of the early Dean innovators and one of my favorite bloggers, went on to run Barack Obama's blog during his 2004 U.S. Senate campaign. After Obama won, Klau went to work for FeedBurner, a company that makes it easy for bloggers and others to use RSS (really simple syndication) feeds, publicizing and optimizing them so that more people choose to receive their content. The technology was so leading-edge that Google bought FeedBurner for a reported $100 million. Rick and I had first met at a Dean rally in Chicago in 2003. Now I was speaking on Google's Mountain View, California, campus at the invitation of the latest member of Google's team: Rick Klau.

Like Rick, many of the Dean pioneers kept on pioneering. Ginny Hunt, who had helped start Generation Dean and sparked the campaign's online youth movement, was working at Google now too as the company's election programs manager. Nicco Mele, a partner in EchoDitto, also worked on Barack Obama's U.S. Senate campaign; the other real Dean campaign spin-off, Blue State Digital, would rebuild the Democratic National Committee's web site. With Howard Dean as its new chairman, the party was embarking on a new fifty-state strategy, and Blue State was building the web tools to help the DNC get there. Then they wound up with a hot one: building the web site for Barack Obama's presidential campaign.

I looked around the political landscape like a beaming father. The Dean Campaign was everywhere. Karl "with a K" Frisch, a key member of the Dean team, was over at Media Matters, a progressive site aimed at keeping the mainstream media honest. David Bender, who had recruited me into the Kennedy campaign nearly thirty years ago, and who'd been there with me from the start in the Dean campaign, was now a leading progressive voice on the Air America radio network. Michael Silberman, who had run the Dean campaign's Meetup strategy, was now at 1Sky working to organize a nationwide effort to confront climate change. Matt Gross, who'd launched the Call for Action blog for Howard Dean—the first presidential campaign blog in history—and Japhet Els, one of the youngest Dean staffers in 2004, were helping John Edwards's 2008 campaign team get their Internet efforts off the ground. Jim Brayton, one of the most innovative—and sanest—members of the Dean campaign, helped get Barack Obama's Internet campaign going strong, before proving his sanity again and leaving the campaign to be with his family in time to prepare for their new baby. And Joe Rospars, one of the Dean campaign's top bloggers, took a leave from Blue State Digital to help Barack Obama full-time, overseeing the campaign's entire Internet and new media communications.

I remember going to my first YearlyKos convention, a meeting for progressive bloggers, and being stunned at how many of us there were, and . . . well, shocked at how well organized it was. Even a potential presidential candidate or two showed up, hoping to drum up some support among the netroots. Governor Mark Warner of Virginia blew away all the competition in 2006 with a party that no self-respecting blogger could believe, complete with chocolate fountains and rollercoaster rides off the top of the Stratosphere Hotel and Casino in Vegas. But that was the point: bloggers had been laughed at by too many in the party in 2004. Now Warner and other candidates were treating them with some frickin' respect.

By the time 2006 was over, at least two U.S. senators—Tester in Montana and Jim Webb in Virginia—owed a large part of their victories to early support from online activists. Webb also got a YouTube boost when Senator George Allen had his "macaca" moment, when he was caught on tape berating a member of Webb's staff with a racial slur.

The comment got caught in the echo chamber between cable news and YouTube—and its echo didn't subside until Allen was no longer a U.S. senator. Now Virginia's junior senator would deliver the Democratic Party response to George Bush's 2007 State of the Union address while a growing online community cheered him on.

Like it or not, Joe Lieberman was no longer in the Democratic Party, courtesy of the progressive netroots. Ned Lamont had won Connecticut's Democratic Senate nomination by defeating Lieberman in the primary, but Lieberman turned around and ran as an independent in the general election as a political maneuver to hold on to his Senate seat—and it worked. Lieberman and much of the party establishment blamed the netroots for the upheaval in the party, but they didn't seem to understand that the netroots had only raised money and helped organize Lamont's campaign; it was the citizens of Connecticut who cast the votes to defeat Lieberman in the Democratic primary.

The 2006 race also saw the emergence of new members of Congress who had raised money on ActBlue.com, a new netroots site designed to raise resources for campaigns that often weren't on Washington's or the party establishment's radar. With just a few clicks you could make a donation to multiple candidates—and only have to fill out the required forms once! *Really cool*, as Nicco would say. Candidates were actually being recruited to run for office and funded by the netroots, and winning—even when Washington thought it had a better candidate. The establishment was used to finding a candidate who could raise big checks—or who could write one big check to fund his or her own campaign. Now the game was changing: thousands of individuals were making small donations with a click or two, providing the campaign sweat that gave birth to dozens of winning campaigns in 2006—enough for the Democrats to take a majority in the House. Nancy Pelosi was the new speaker of the House of Representatives, and no big party committee in Washington would ever ignore the netroots again.

DeanLink and DeanSpace paled in comparison to MySpace and Facebook and the millions and millions who were members of social networks as 2008 approached. Where Dean had had a photo gallery, now Flickr.com allowed millions to post photos, "tagging" them so that if you searched your favorite candidate's name you could view every picture taken at a rally and posted just a few hours before. Our primitive text messaging network on Upoc.com may have been visionary at the time, but just four short years later it was nothing compared to Twitter, the hot new mobile social texting network.

The Dean campaign had been the Wright brothers of this new kind of politics. In the next four years, however, the world would skip Boeing, Mercury, Gemini, and everything in between. By the start of 2008, the Apollo project of a new kind of politics was moving toward its launch pad.

But the tools weren't the big game changer between the 2004 and the 2008 election cycle; the people were. The reason I knew we were approaching the tipping point was the simple fact that the number of people using these tools had been increasing every day. And a hell of a lot more people—millions more—were going online; millions more had broadband connections; millions more were transacting, blogging, and networking than we ever imagined during Howard Dean's heyday. The bottom wasn't just "getting it"; the bottom was growing, peer-to-peer and friend-to-friend, gaining strength every day and taking power away from those at the top, whether they liked it or not.

The implications for the 2008 election cycle ran through my mind. I told anyone who would listen that before it was all over one of the 2008 candidates would raise more than half a billion dollars. I'd gotten used to establishment types in my party looking at me like I was wearing a tinfoil hat. But I knew I was right. I wanted to shake the top of every presidential campaign: *You better get it!* Because this time there'd be no mercy for anyone thinking big money and top-down control could stop what was coming.

The 2008 campaign would be the second Internet-savvy, bottom-up campaign cycle, and it would end the doubts forever. The money, the crowds, the number of volunteers would dwarf the Dean campaign— just as whatever numbers the 2008 campaign puts on the board will pale in comparison to the numbers put up by some campaign in 2012 or 2016.

It's the network, stupid, and it's growing bigger and stronger every day. And I was sure it was big enough and strong enough to topple top-down politics in 2008.

I had toiled in the hell of transactional politics for three decades— a time when television dominated everything. I had worked for some incredible people. But we'd all been trapped by that damn box and the money it sucked in—and, yes, by the lies that too often spilled out of it. The truth is, I *hated* transactional politics. I despised the selfish, *what's*

in it for me politics, and I'd lost faith in a political class of elites ready
to promise anything for a vote. Now, finally, all that was on the verge
of being upended. It looked as though the people might be ready and
able to declare transactional politics dead, and start working together
for the common good again.

And where was I? Sitting at home, watching *Law & Order* reruns
and feeling sorry for myself.

But I couldn't stop thinking about doing it again.

In late 2006 and early 2007, I met with every campaign who wanted
my advice, or who just wanted to ask me what had "gone wrong" in
2004, so they could avoid our mistakes. I met with some key people
in the Clinton campaign, her campaign manager and some others, but
when I went through my change rap I was met with vacant stares:

"Hillary Clinton should run as a change candidate," I said. "She
should say, 'I am going to do this differently from the way it's been
done before. I will not take money from Washington lobbyists or PACs.
I believe that there are five million women in America who are willing
to contribute $100 each to change this country and our politics. Every
woman within the sound of my voice knows that we're not the status
quo; we are *change*. We have a different perspective and a different way
of looking at things—on why it's so important to educate our children,
to protect a planet that's in peril, and to end war. Not just this war, but
all war. My name is Hillary Rodham Clinton and I will tell you this:
We didn't start the fire, but we are damn sure going to put it out.' That's
what she should say."

Can you say tinfoil hat? They did the math right there in front of
me. Five million women giving $100 would be half a billion dollars. The
campaign manager looked up at me and said, "That isn't possible." They
didn't get it. I knew the campaign had hired Peter Daou, a brilliant
bottom-up Internet strategist, to run Clinton's online effort. But all the
bottom-up brilliance in the world doesn't matter if the powers that be
still want to run a top-down campaign.

A few weeks later, I was in New York participating in some panel
discussion with Mark Penn, Clinton's chief strategist. We were asked
what the impact of the Internet would be in 2008. Penn jumped to answer the question, saying that the Internet wouldn't have any impact in
2008 because (I kid you not), "It is composed of too small a group of
Americans who are doing nothing but talking to each other." I nearly
fell out of my chair. By the end of the panel, I knew Penn was trapped

in old-school thinking. Either the Clinton campaign was doomed—or it would be the last top-down campaign to win the Democratic nomination.

At some point, Kris Balderston, Clinton's deputy chief of staff; Howard Wolfson, her campaign's communications chief; and Jay Carson, her traveling spokesperson, all talked with me. To their credit, they all believed she could win by running a bottom-up campaign, and they were really open to my ideas and those of others. But I was certain there was no way Penn would ever allow it—and, besides, I had that one-horned goat at home I had to take care of.

By early 2007, I saw something new on the horizon: Barack Obama was beginning to rock.

It was clear to me that *he knew*—he understood the moment. He had been a community organizer. He was Mr. Bottom Up.

You had to be blind not to see the buzz Barack Obama was generating online and throughout the country. He had something Howard Dean never had: plenty of tough political operatives who'd been through the hard knocks of a presidential campaign before. You could count the senior staff members in Dean's campaign who'd worked in presidential politics on two, maybe three fingers. Obama, on the other hand, would have plenty of experience to help organize his bottom-up candidacy as it was catching fire.

David Axelrod, his chief strategist, is one of the best and most decent people I have ever met in this game. I guest-lectured in his class at Northwestern University after the Dean campaign, and he was kind enough to make this book required reading in his class. I knew Ax and I believed in the same things; he hated transactional politics as much as I did. Barack Obama was going to run a transformational campaign—that I knew for sure.

When I met with Axelrod in Chicago, it was almost embarrassing: he had thought out everything. I went through the five-million-people-giving-$100 talk; Ax just looked up and said, "Yeah, I read that somewhere. I think it was toward the back of your book." I admired everything he said, but I also knew that Obama was trying to keep the core of his campaign staffed with people he had worked with and knew. I was excited about where his candidacy might go, but I also wasn't sure whether his policies were as bold as those of Edwards—the other candidate who was mounting a real bottom-up campaign.

All of this and more was whirling through my head in the weeks after I hightailed it out of that meeting with John and Elizabeth Edwards. I had in one way or another talked to all three of the top campaigns for my party's nomination, and I'd ended up not working for any of them. 2008 was shaping up to be the historic election in which bottom-up politics would finally triumph over the old transactional politics—and where was Joe Trippi going to be? Sitting on the sidelines whining about his neuropathy.

What was running through my mind? *I think I'm going to puke.*

Then it happened: CNN, MSNBC, Fox, breaking news: John and Elizabeth Edwards were holding a press conference. Ben Smith of Politico.com was reporting that Elizabeth Edwards was suffering a recurrence of her breast cancer and that John Edwards would be quitting the race.

I hadn't talked to them in the three weeks since our meeting, so I watched like everyone else as the commentators—who knew nothing—commented anyway. Waiting for the press conference, I thought about the conversation I'd had with the Edwardses that day. How completely Elizabeth got the Internet, and they both got the idea of bottom-up transformation. I remembered how real and powerful their commitment to universal health care was. And how John Edwards had talked so authentically about his own upbringing and how his life's cause was ending poverty. I also remembered how gracious Elizabeth had been to me that day, so focused and concerned with what I was going through after all she'd gone through herself.

And now this.

I said a silent prayer for Elizabeth Edwards and hoped the rumors weren't true. I remember leaning toward the TV just as the press conference was about to start and muttering out loud to myself, "Please don't get out."

Then the news: It was true, Elizabeth Edwards had had a recurrence of her breast cancer. But then the news I *didn't* expect: That didn't matter. They were staying in. They were going to continue to make a difference and not let cancer win.

I turned and picked up the phone. I wasn't done yet either.

I joined the Edwards campaign on April 16, 2007. On the previous day, the Federal Election Commission reports detailing each campaign's fund-raising and expenditures were made public. So the first question

I was asked by a reporter wasn't, "Hey, how's it feel to be back in it?" Or, "Why Edwards? What do you think he offers the country that the others don't?"

Instead it was, "What can you tell me about the $400 haircut?"

Great. I hadn't been back in for a full day and already I was being reminded of the bullshit that would make any sane person stay the hell out. The country was at war. The globe was on the verge of a climate crisis. Millions were living in poverty; millions more in America had no health care or couldn't afford it any longer; and our working people were being pressured economically and hurt even further by trade deals that increased the profits of multinational corporations while hemorrhaging blue-collar jobs here at home. John Edwards by far had the boldest proposals to deal with these challenges—but all the press was interested in was two $400 haircuts that had shown up on his FEC report.

It was a return to the failed politics of our time, where a rival campaign knows you've got 'em beat on policy, so they find some "gotcha" crap in a finance report and feed it to the press, fanning the fire so high that the reporters actually think they've got something.

In our case, an advance team asking someone to come to a hotel to cut the candidate's hair between campaign events, and then using the wrong credit card to pay for it, would unwittingly create the grounds for near-scandal-level coverage. For months, the *New York Times* was hard-pressed to write a story about John Edwards that didn't reference his haircuts. One of the rival campaigns was insisting (off the record, of course) that the $400 haircuts were proof of John Edwards's hypocrisy, evidence that he was so well off he couldn't possibly really care about poor people. As though you had to be poor to care about the poor. As though the price of a haircut had a damn thing to do with what it took to be president of the United States—with ending the war, ending poverty, passing universal health care, or fighting for policies that helped working people.

It was a sign of how shallow our politics had become that the press was actually rewarding this bullshit: they *admired* campaigns for their ability to distract and transact. John Edwards had been the first candidate to propose a universal health care plan; the Clinton campaign was actually making noises about not presenting a health care plan at all until she was president—claiming that giving away too many details might ruin chances of passing anything significant. Why cover *that*?

Why report the boring details of meeting the nation's challenges when there was all this important stuff about haircuts to report?

What campaign would have an interest in pushing this garbage? Maybe the campaign that didn't have a heath care plan?

Only a cynic could miss the irony. Forty-seven million Americans didn't have health care. More than thirty million were living in poverty, many of them children, in the richest nation on earth. But none of *that* was a scandal—not to the reporters, at least.

The rival campaign was pushing the idea that John Edwards was using his "supposed" focus on ending poverty in America as a cynical way to get votes. Which was ludicrous on its face: as anyone with a political bone in his body could tell you, there are no votes to be gotten by calling for an end to poverty in our country. If ending poverty were a great vote-getter, John Edwards wouldn't have been the only candidate talking about it. He wouldn't have been the only candidate repeatedly putting the Ninth Ward of New Orleans on his schedule against the advice of many of his campaign staff. (Oh, we believed in ending poverty too. I loved him for it. But the votes he needed were in a place called Iowa.)

Stopping the references to those haircuts was like trying to plug holes in a leaking dam. Every time we thought we had seen the last reference, another one would pop up.

Until our YouTube video stopped them cold.

The CNN/YouTube Democratic debate was coming up, and for a few weeks Jonathan Prince, our deputy campaign manager, kept yammering at me that we needed to do a YouTube video that somehow addressed the "hair thing." What would be in it? It would show *other* people's hair—that was all we knew.

Prince had been with Edwards in 2004 when I was with Dean, and we were just starting to understand each other. The man was a walking Starbucks coffee mug: once when Starbucks stock went down 11 percent, I thought Prince must have stopped drinking the stuff. One day, we were back in Starbucks—he'd dragged me there for the eighth time that day—when I brought up an idea that could make the whole thing work: we could build the video around the title song from the 1960s musical *Hair*. We called Jordan Pietzsch, a bright young associate in my firm, talked him through the idea, and told him to see what he could come up with. A day later Jordan had delivered a frickin' masterpiece— and he'd cut the whole damn thing on his Mac.

Jordan started the whole thing off with a shot of Edwards's hair, followed by a parade of fabulous shots of people's hair, just like Jonathan had wanted. But then came the amazing stuff: troops and Iraqis running for cover, victims of Katrina in a New Orleans intersection with the word "Help!" scrawled in chalk across the entire street, women in California working for health care reform on a street corner holding a sign that read "honk for health care."

The piece was edgy and raw and in-your-face—but it needed one last tweak. I had Jordan fade a screen up at the end that asked the simple question, "What really matters?" And Jonathan finished it off with one last screen reading "You Choose," to play into YouTube's YouChoose election year theme. When the video went up during the CNN/YouTube debate it was different from anything any other campaign had done. It was dangerous: we were raising the whole hair thing again, rubbing it in the press corps' faces. You couldn't watch the thing without being disgusted at their fixation with hair in the face of all these other problems.

The video got rave reviews: hundreds of thousands watched it on YouTube, the bloggers got blogging about it, and soon hundreds of thousands of dollars were being contributed to our campaign—and the kicker was, it worked. The $400 haircut issue never really came up again. There had been other YouTube videos, like the "macaca" moment, that had *started* campaign fires; this was the first one anyone could remember that had helped put one out.

But our "Hair" video was really just another example, and not nearly the most powerful one, of how social and viral media as well as social networks and the millions that were using them were helping to transform our politics in 2008.

Back in February, before I joined the Edwards campaign, Farouk Olu Aregbe, a twenty-six-year-old in Columbia, Missouri, was so excited about Barack Obama's candidacy that he started a group on Facebook.com named "One Million Strong for Barack." I didn't know who the hell the guy was—*no one* knew who the hell the guy was—but by the time I joined Edwards in April, just six weeks later, more than four hundred thousand people had joined Farouk's quest to bring one million supporters into Obama's camp.

More than four hundred thousand in just six weeks. *Oh, shit.*

The Obama for America campaign didn't know who Farouk was either. He was just one guy out there in Missouri who used the social networking site to create a forest fire of supporters. Each supporter asked his or her own personal network to join the cause, and each new supporter asked everyone in his or her personal network—mothers, brothers, coworkers, and friends—to do the same. Before it was over, Chris Hughes, one of Facebook's founders, signed on—not just to join the cause but to become Barack Obama's online social networks director. Obama was attracting some heavy-duty talent and it wasn't all on Facebook. The actor Tom Hanks announced his support for Obama on his MySpace page late in the primary campaign.

I had a reporter ask me if it really mattered that so many people were performing such a simple task as joining a candidate's support group on Facebook or MySpace. I told him that most top-down presidential campaigns would be thrilled to get fifty thousand people to do something as simple as putting a bumper sticker on their car. But here's the problem: Let's say I see my best friend with a bumper sticker on his car, and I want to put one on mine. What do I do? He doesn't have one to give me, and even if he did, I wouldn't have any to give *my* friends. Bumper stickers are essentially a very slow way for campaigns to reach out to people *one at a time*.

Instead, on Facebook, one savvy Obama supporter had created a group of *hundreds of thousands* of people, who would all be able to use the new bottom-up applications to make get-out-the-vote calls from their homes, turn out neighbors to meet the candidate, create huge local crowds, and drum up contributions that would help break online fundraising records—which were set by some guy named Howard Dean.

And, by the way, every one of them could get their own bumper sticker with one click.

And none of this was organized by Obama's campaign. They had nothing to do with it. It was the network, stupid, and the network was growing every day.

In the Edwards campaign, we knew what we were up against: two eight-hundred-pound gorillas and a press that was fixated on both of them. At the YearlyKos debate, John Edwards repeatedly called on Hillary Clinton and Barack Obama to join him in calling for the entire Democratic Party to stop taking money from lobbyists in Washington. One exchange, triggered by our guy, prompted Clinton to answer with her now-infamous "Lobbyists represent real Americans" gaffe as an excuse for refusing to join us. The next day's headlines? "Clinton, Obama Disagree on Lobbyists." John Edwards was still the only candidate calling on the party to stop taking the money; we'd scored the point, but somehow it got lost in the coverage again.

It was clear to us that Hillary Clinton's top-down campaign was going to corner 80 percent of the market on $2,300 dinner ticket contributions. But it would be a long time, if ever, before she became a threat online. Announcing her candidacy online had been a great idea—but she was reading a script on a teleprompter, and the whole thing had the feel of an artificial television broadcast. She kept taking PAC and lobbyist money, even though she didn't have to, and her campaign didn't seem to understand that if she would say she'd stop taking it she would raise a lot more from small donors online than she'd ever give up in special interest money and the claims they'd make on her White House. And the one good Internet moment for her campaign—the *Sopranos* parody that introduced her campaign's winning theme song—was exactly what a top-down campaign would think a YouTube video should be: scripted and safe. The only reason it was cool was that the Clintons had made an online video; that's why it got a lot of buzz with the press, but there was nothing real or authentic about it. The campaign just wasn't capable of taking chances by being real. Which was a shame, because every time you saw Hillary Clinton be real, like when she choked up in New Hampshire, people really responded to her.

I always thought, deep down, that at some point the top of the Clinton campaign would wake up. So I admit I was stunned to watch them stubbornly persist with the old command-and-control style campaign, eyes closed to the possibility that the American people were smarter than they were. But that's what the Clinton campaign was doing.

Terry McAuliffe, the Clinton campaign's chairman, was building an old-style network of powerful big money contributors, beating the performance of any previous Democratic campaign. Even before the race really began, they'd built something like a two-hundred-delegate lead with superdelegates. Their strategy was to end it early, to win in Iowa, New Hampshire, and Nevada, then finish off anyone left standing in South Carolina. And they *knew* that was how the whole thing was going down. It would be just like Mark Penn had said: no need to worry about the Internet, that tiny "group of Americans doing nothing but talking to each other."

Meanwhile, though, the second bottom-up, Internet-savvy campaign was building strength, fueling its tanks and readying to launch with a network exponentially larger and more powerful than the one we had in 2004. This wasn't going to be some flimsy contraption that proved the aerodynamics of a new politics. This thing was going to challenge the supremacy of top-down politics and if it got off the pad might just carry the right candidate to the White House.

In the early days, the Edwards campaign and the Obama campaign both did really well online. Obama was dominating the social networks and clearly setting a record pace with online fund-raising, but John Edwards's policies were generating a lot more interest online than Obama's, and we led most of the online community polls. We weren't breaking any records on the fund-raising side, but our online fund-raising was running ahead of Howard Dean's 2004 pace.

But we weren't stupid. We knew we really only had one chance. No matter what we did, the press kept focusing on the race between Clinton and Obama. That's not a complaint; it's just the way it was. So our only real chance was to beat Clinton and Obama in Iowa. We had to stun everyone by beating not just one of the eight-hundred-pound gorillas but *both* of them. If we could take a media explosion created with an Iowa victory, and mix it with our strong online presence and John's bold ideas, maybe, just maybe, that rocket would have John Edwards's name on it.

You didn't have to pick up the *New York Times* or the *National Journal* or even the *Des Moines Register* to know how steep the odds of beating two candidates marching into history were. Our strategy was realistic but tough. And I was proud to face the challenge with a great team, an incredibly smart and indefatigable candidate, and his amazing wife.

I was sitting in the campaign headquarters in Chapel Hill when Ann Coulter, the right-wing nut job, was about to appear on *Hardball with Chris Matthews* on MSNBC, just a day after she'd trashed John Edwards with a particularly nasty personal attack. Elizabeth Edwards got on the line with me, told me she wanted to call in to the show and "talk" to Coulter.

Oh shit! And I gave her the number!

But it worked. Elizabeth handled her side of the exchange with politeness and charm; Ann Coulter, on the other hand, proved once more what a craven bottom-feeder she is. And as a result our supporters rallied around Elizabeth and her call for more civil political discourse, contributing another $300,000 online to our campaign. Clearly the online support was there for us if we could light the rocket with a win in Iowa.

Then came the debate in Philadelphia. Hillary Clinton screwed up big-time, answering just about every question with at least two different answers. We asked Jordan to take a look at the debate tape and see what he could do with it for YouTube. What he produced roared around the Net: A video that showed Hillary saying she'd bring our troops home from Iraq, and then saying she'd leave them in. That showed her saying she had a very specific plan to reform Social Security, and then showed her saying she wasn't going to propose a specific plan on Social Security. And the finale was Hillary Clinton on four different screens giving four different answers to Tim Russert's yes-or-no question on driver's licenses for illegal aliens. And that's when Jordan ended it with a screen fading up saying "That's the Politics of Parsing." The press called it "devastating," and it was. The clip jumped from YouTube to the cable networks, and the echo chamber went into full force. The second this thing hit YouTube it helped raise real doubts about Hillary Clinton's credibility—doubts that have stayed with her through the rest of the campaign. I can't believe Jordan did that!

And there was the time when I squared off against Mark Penn from the Clinton campaign and David Axelrod from Obama on *Hardball.* A senior Clinton official in New Hampshire had gone after Barack Obama because he'd admitted—in his own book, no less—that he'd used drugs in his youth. The national Clinton campaign was denying they had anything to do with the attack. But there was Mark Penn, standing right beside me as we did a two-shot on live TV, raising the smear again even as he tried to deny that his campaign had anything to do with raising it.

"Well, I think we've made clear that the issue related to cocaine use is not something that the campaign was in any way raising, and I think that's been made clear," Penn said.

I couldn't believe what I was hearing. I called Penn out on the attack. "I think he just did it again! He just did it again! Unbelievable. This guy's been filibustering on this. He just said 'cocaine' again."

Then Penn tried to blame it on me. "I think you're saying 'cocaine.' I think you're saying it." Penn was now going so low that everyone could see what he was doing. The entire exchange jumped from MSNBC onto YouTube, and the link was posted on blogs that were read by everyone following the fight for the Democratic nomination online.

In a post on MyDD.com, Todd Beeton summed up how many in the blogosphere viewed the exchange and what it said about the three campaigns. He gave a shout-out to my "fire in the belly," saying that it "reflected perfectly Edwards's populist fighter persona," and noted how David Axelrod's "mellow, above-the-fray style echoed Obama." As for Clinton's representative, he could only ask: "What does Penn's slimy shiftiness say about his candidate?" Just another way that YouTube and the Internet would help those moments of truth in the campaign live on, capturing those unflattering glimpses of the Clinton campaign and its do-anything, say-anything tactics.

Amy Rubin, one of Edwards's online organizers, kept pestering us about Eventful.com, a new site where people could go to ask or "demand" that their favorite band or personality visit their town. Amy's idea was to use the site in a contest, giving people across the country the chance to "demand" that John Edwards campaign in their town; the town that organized the most votes would get him for a day. We launched the contest and watched as big cities like Los Angeles and Seattle took the lead, but we were in for a big surprise when we found out who the winner was.

Tracy Russo, our chief online organizer, came into my office that day. "It's Columbus, Kentucky!"

"Where the hell is Columbus, Kentucky?" I asked.

Columbus, it turns out, is a town of 229 people. The closest Mc-Donald's is more than fifty miles away. It was a small town in a red state—the kind of place Democrats had to win in November, but that never saw a Democratic candidate in the primaries. But somehow the good people of Columbus had convinced enough friends and family members in other towns and cities across America to vote for it—and now John Edwards was headed there, with the national press in tow. It would prove a free press bonanza, with cable news coverage and stories in newspapers around the country. But the real surprise was the crowd: people came from all over, swelling little Columbus to more than two thousand people. For one bright sunny day of campaigning on the banks of the Mississippi, we really were all in this together.

The top of the Clinton campaign had ceded the online community to us and Obama without ever really competing for it. We were fighting for it with everything we had. But it didn't take long to see that we were losing ground to Obama. The more the press focused on Clinton and Obama, the more Obama benefited and grew online. It was something

I had learned in 2004: as much as the Internet is going to free us from that damn box, today the two still work in tandem. Obama had a big, hopeful message that was capturing people's attention, and with it the attention of the press. In the Dean campaign, we'd understood that relationship better than anyone. Now Obama was the benefactor. Hillary Clinton's campaign seemed clueless; they seemed to be counting on the idea that her big money would take care of whatever the hell this was.

Where John and Elizabeth and our team excelled was in presenting new bold policies that pointed in the direction John Edwards believed our party and our nation must go. And he never gave an inch. Pushing a real plan with a time line to get our troops out of Iraq when neither Clinton nor Obama wanted to go there—until they both finally moved. It was like watching a giant lever: with John Edwards as the rock, the people of Iowa provided the weight that shifted where the other candidates stood. And it happened on all the important issues: on Iraq, on the special interests that have wrecked Washington, and on the trade deals that have wrecked middle America, they moved. Sometimes John Edwards was the rock and Obama was the weight that moved Clinton. Like on health care, where John Edwards came forward with his universal health care plan. Clinton was hoping to avoid committing to a specific plan, but then Obama came out with his own health care plan. Soon Edwards was out there telling the people that his health care plan was better than Obama's, but that "at least Barack has a plan." Finally, Clinton caved and produced one. Throughout 2008, on the policies that mattered most, John Edwards was always the rock. He staked out the position and held his ground and made them move.

Later, as I watched the campaign throughout the spring of 2008, as Obama and Clinton campaigned through Ohio, Pennsylvania, Indiana, and North Carolina, I was struck by how much they both sounded like John Edwards.

They even started talking about poverty. John Edwards got their commitment on that issue too—challenging them both to put ending poverty at the forefront if they won the presidency.

Unfortunately, it was a challenge he issued as he got out of the race.

What happened? We didn't win Iowa. We beat the Clinton campaign, despite its millions of dollars and more than 400 organizers on the ground in the Hawkeye State, compared to our 176. But Barack

Obama outspent us by even more, and beat us by about seven percentage points. Obama received 38 percent to Edwards's 31 percent and Clinton's 30 percent. Though it wouldn't become clear for a state or two, Obama had effectively run us out of the race. Close rarely matters in politics. And the press didn't miss a beat: suddenly the story of New Hampshire was all about whether Hillary Clinton could pull off a comeback. The press corps became fixated almost exclusively on the Clinton vs. Obama story, and we fell even further out of the picture—though we tried valiantly to stay relevant and force them to cover us.

At one point, as we traveled around South Carolina in a bus, with maybe a handful of press traveling with us, Edwards turned to me and asked, "What do I have to do? Set myself on fire?" We both knew the answer. And even that wouldn't have worked—or I would have tried it.

Our campaign ended the way it started, in the Ninth Ward of New Orleans, with John Edwards once more trying to bring attention to poverty, and our nation's failure to come to the aid of its people. "I began my presidential campaign here to remind the country that we, as citizens and as a government, have a moral responsibility to each other, and what we do together matters," he said. "We must do better, if we want to live up to the great promise of this country that we all love so much."

It was worth it. I would have done it all over again. I was proud to have ridden around on that campaign bus with John Edwards, even in those moments when those bees were stinging away with every bounce and bump on America's highways. I believed we had that moral responsibility to each other. And nothing made me prouder than helping John and Elizabeth Edwards lead our party and the country in that direction again. It was the transformational call the nation needed to hear—a call to throw transactional politics aside and do what was required of us as Americans for the good of the nation and each other. Sometimes that's just how it works: you fight to point things as close to the right direction as you can—and then it's time to get out of the way.

As John Edwards so eloquently put it that day in New Orleans as he closed out his campaign, "It's time for me to step aside so that history can blaze its path. We do not know who will take the final steps to 1600 Pennsylvania Avenue, but what we do know is that our Democratic Party will make history."

I'm not going to mince words here: John McCain, what the hell happened to you?

The McCain campaign of 2000 was the frickin' campaign that scared me into thinking the GOP was conjuring up some new tech/new media mojo on the Internet that would put my party behind the eight ball for another decade or two, just as they'd done with direct mail and talk radio. The forty thousand people who had signed on with McCain on the Internet, and the millions of dollars those people sent him the day after he beat Bush in New Hampshire, had gotten my attention. That was the whole reason I decided to shed my old top-down politics clothes and embrace the new way.

But 2008 was a different story. It was as if McCain, and just about every other Republican for president, had just stood still for eight years!

David All, and some other GOP online organizers, approached me from time to time at Web 2.0 gabfests to ask me to do an online interview just so they could post it and encourage others Republicans to see the light. I've always thought that our democracy only gets stronger if more citizens are active and involved and part of the process; I think many of our biggest problems today have come about because most of us weren't paying attention until it was too late. So, in the spirit of bipartisanship and the Marquess of Queensberry rules of fairness, I always did those interviews and challenged the other side to help build a more participatory democracy.

Result? Nada. By 2008, the onetime maverick McCain had gone GOP establishment, and when it came to bottom-up campaigning, he was having none of it.

The GOP did have another guy, Ron Paul, who was trying to build an online community with a lot of success. His campaign supporters were out there dropping "fund-raising bombs"—choosing a specific day and time when they'd all rally together to raise a set amount of money. People in the Republican Party laughed at Paul's supporters—until, on November 7, 2007, their "fund-raising bomb" dropped $4 million into his campaign in one day. I took notice, and the GOP establishment would be wise to take notice too. They're going to need all the help they can get in the 2008 general election.

Hillary Clinton's campaign would pay a heavy price for its failure to recognize just how much everything has changed since 1992 and 1996, the years when most of her campaign operatives had last been involved in a presidential campaign. They'd built a campaign insisting that everything remain under the core staff's complete control—with all the creative output coming from within the walls of the campaign. It was a campaign that would rely on maxed-out donors and build a top-down fund-raising network the likes of which the Democratic party had never seen. On policy it kept its options open, and when it came to cutting up its competition it had the sharpest knives in the drawer—so good that they bragged about it as a reason they should face the Republicans. It was, by just about every measure, the greatest top-down campaign in history.

But the Clinton campaign's old-school approach would create the two blunders that would leave it vulnerable to the second bottom-up Internet political campaign in history.

The first was that dependence on donors who would give it $2,300 checks, the maximum allowed by law. The problem for the Clinton campaign was that they could never go back to these contributors again. Once a contributor had maxed out, all the Clinton campaign could do was ask them to find a friend to contribute $2,300. After a while, that game just wasn't going to work. The Clinton campaign, it seemed, had missed the lessons of 2004. The Dean campaign had out-raised everyone in all of 2003 because it was able to go back to its hundreds of thousands of small online contributors again and again. In contrast, John Kerry, who ran his campaign on the top-down model, was only able to compete with Dean down the stretch by loaning his own campaign more than $6 million.

Now it was the Obama campaign—working with a much bigger network than Dean—getting its supporters to get their friends to join his campaign, and then asking them all, millions of them, to give what they could afford. Now all these people were giving $25 or $45 to the cause, and the Obama campaign was able to go back to the well again and again, just as we had for Howard Dean's campaign. But Barack Obama struck a bottom-up gusher and it ran a hell of a lot deeper.

Throughout the race, Hillary Clinton's campaign was constantly running out of money. The campaign's big money network went dry again and again. And Hillary, like John Kerry, would have to loan her campaign money just to keep it afloat—more than $11 million dollars.

The second bottom-up campaign in history was out-raising the greatest top-down campaign in history. It was the network, stupid. And Barack Obama's was bigger.

But the blunder that may go down as one of the biggest of all time—the blunder that may have delivered the Democratic nomination to Barack Obama—was the Clinton campaign's decision to bypass most of the states that were holding party caucuses. The Obama campaign used its online organizing tools to plug its hundreds of thousands of decentralized volunteers into its professionally run caucus organizations, and that helped it break the Clinton campaign's back in caucus state after caucus state—giving Obama huge wins and the delegates he needed to pull ahead in the delegate hunt for the nomination. Never, in thirty years of doing this, have I seen a bigger mistake. To be sure, from the start Obama had a bottom-up advantage in these states, where grass-roots activity on the ground can pay off big. But the Clinton campaign downright abandoned the field; they just didn't play. They had no organizing structure in the caucus states, because in a top-down campaign you have to pay people to build one—and the Clinton campaign was out of money. The Obama campaign's hundreds of thousands of decentralized caucus volunteers organized his caucus victories for free—and most of them were also contributing money to his campaign. Clinton just couldn't compete.

Barack Obama was truly a community organizer at heart. And he was within a beat or two of taking the nomination from a campaign that never saw him coming.

And it wasn't just the Obama campaign's small-dollar fund-raising or the hundreds of thousands, maybe millions, of volunteers who joined his cause that did it. It was the sheer creativity of the American people who went to work for his campaign—people who believed that what they did mattered.

The Clinton campaign made the tragic mistake of telegraphing that the campaign had it all figured out, that the only thing that mattered was what happened within the walls of the campaign.

The Obama campaign, on the other hand, had taken a lesson from the flimsy-contraption days of the Dean campaign. They used the better tools they had and the much stronger organizational structure they were able to build. But they also learned to keep any walls they built low, and made sure the campaign's supporters knew their creative ideas were welcome.

When Obama gave his "Yes We Can" speech, supporters created mashups of images and music to the sounds of the candidate's voice. will.i.am of the Black Eyed Peas made a version that took the nation by storm, with more than 13 million people watching his YouTube take on Obama's words in a matter of weeks. To give you an idea of how big an audience that is, compare it with the 2.5 million Americans who watched the CNN/YouTube debate. Or the combined 20 million or so Americans who watch network news on ABC, CBS, and NBC on an average evening. And will.i.am didn't even have the only version of that mashup: millions of Americans were taking part, creating and watching and signing up. *Yes We Can. Yes We Can.* And millions of people believed they could.

One day in February, after John Edwards left the race, I woke up in New York City to something I never thought I'd see. In my new role as a CBS News political consultant, I'd been up late the evening before, working with Katie Couric and the *CBS Evening News* team in their coverage of the Super Tuesday results. Now I couldn't believe what I was hearing: the greatest top-down political campaign in history was dead.

What had I seen? Hillary Clinton, uttering the words I thought I'd never hear her say on the air: "Please help me by going to www dot Hillary Clinton dot com."

The only way I knew it was really her was that she said the "www" prefix out loud. No one uses that anymore.

Call it desperation, or simply a late awakening, but the greatest top-down campaign in history was going bottom-up—and in a big way. By the end of February 2008, Hillary Clinton would raise $35 million, almost all of it online. She'd finally called on her own community to come to her aid, and they had responded.

Hillary Clinton may have "found her voice" in New Hampshire, but she found the People in February 2008. Her efforts were clunky at first, but at least she was really making them. And I didn't care if desperation had caused the discovery or not—John Edwards was right. One way or another, the Democrats would make history. And now, no matter who won, we'd be running a bottom-up campaign that believed in the power and energy of the American people.

I'm pretty sure no one will make the mistake of running a top-down campaign for president again. (Cynics, reach for your barf bags.)

For thirty years I have done this because I love my country. Because I want to make a difference. For years, I wrestled with the question: what will change the lives of people more, technology or politics? Somewhere along the way, though, I realized that the answer was neither. It was people who would change things for themselves and each other; technology and politics were just tools we could use to work together to improve our lives and build a better future. What I learned in those thirty years was that our politics had failed us, and that people were using technology to join each other in communities as a response to how politicians and politics had conspired to tear the American community apart.

As I write this, in the early summer of 2008, I don't know who will win the presidency. Anyone who says they do isn't dealing with the reality of Trippi's Rule: all it takes is that six-second sound byte that lasts forever, and even the best campaign can be taken down.

In the meantime, though, I know this: Barack Obama and his campaign have showed us all what is possible—what a campaign of connected Americans is capable of doing. They've showed us how the rules are still changing, pushing us closer and closer to a politics where the people matter more than big money, and where we slam the door on the selfish transactional politics that got us into the mess we're in, instead embracing a new transformational politics that recognizes that we're all in this together. Top-down politics as we know it is in its dying days— as is just about everything that's top-down and only top-down in our society.

In 2004, we saw the Wright brothers of bottom-up politics start a new era in campaign history. I was honored to help lead that group of pioneers. In 2008, the astronauts of this new politics are walking to the launchpad. And the general election they're planning will be like nothing our country has ever seen.

I still dream of a presidential candidate who will place his or her fate in the hands of the American people, by limiting contributions to $100 or maybe $250 each. This would allow us to end the power of the monied "factions" our founders feared so much, and return the presidency and our democracy more fully to the people. That dream may still come true in the general election of 2008, but if it doesn't happen now, I believe it will in the next election.

A FINAL NOTE

You may have wondered about that phrase "the overthrow of everything" that's part of this book's subtitle. I mean it. You can't understand how radically the Internet, social networks, and social and viral media are changing our politics, and somehow still believe that your institution, your company, your country will somehow be immune. Everything is on the verge of big, sweeping change—more change than most of us can imagine—and it's coming at a very rapid pace. My advice? Stop denying the inevitable, start adapting now.

Nearly half a century ago, after taking the oath of office, John F. Kennedy turned to a bank of television cameras to give his inaugural address. It was the first time in our history that an inaugural was delivered live on television to an America gathered around its living room sets to see the new president. The combination, in many ways, changed everything. A young president's stirring words challenging the country—not a distant voice on the radio, or a printed speech in the next day's newspaper, but a live address—created a new spirit in the American people. We believed to our core that there really wasn't anything we couldn't do. The television presidency was born on that day.

I believe we're there again. In January 2009, a new president will take the oath of office in a completely different communications environment. For the first time, a new president will be able to use the inaugural address not only to outline the agenda of the executive branch in the first one hundred days, but to issue a call to the American people to join in passing that agenda. Thousands will discuss the new president's words on blogs; many will put them to music, or join them with other images, and create a YouTube video that touches the hearts of people in some different way. Ten million Americans might join the president's call to pass universal health care. And those ten million might use their social networks to get another ten million to join in. On that day in January, the next president of the United States will be standing at the end of the television presidency, and at the beginning of something different. The networked presidency? The interactive presidency? We may not quite know what to call it yet. But it's coming at just the right time. The next president—and the American people—have a lot of work to do. Now, finally, we have the tools to do that work together.

APPENDIX ON THE WEB

For more information about this book, the author, or the Dean for America campaign, go to JoeTrippi.com. There you'll find an Appendix on the Web—with memos, blog posts, and other campaign documents (including the Perfect Storm blog and the Definitional Moment memo), photos, and details on author appearances.

www.changeforamerica.com

ACKNOWLEDGMENTS

F OR THE PAST thirty years, I knew how to get someone elected. I knew
which precinct to walk, which member of Congress or labor leader or
operative could help solve a political problem, what it took to mount a
winning campaign. Strangely to some, I also knew how to get things done
in the tech world. I could talk encrypted chip technology, fiber optics,
and nanotechnology. I could hop a plane for Greece to help elect a prime
minister, then head over to Ireland to meet with a leading Bluetooth chip
company looking for a way to make paper communicate with PCs and television sets. I could be in San Jose making campaign spots to help elect the
city's first Latino Mayor, and the next day, be in a conference room in Silicon Valley looking at ways to make PCs more secure. I made my share of
mistakes, but in the worlds of politics and technology, I at least had some
clue as to what I was doing.

But when I found myself wanting to write this book I had no clue—
no idea at all about how to do this. Tina Brown, Dee Dee Myers, and Lisa
DePaulo listened and believed I had a book in me, and they suggested that
Bob Barnett would be the perfect person to guide me and help me
through the process of getting a proposal together and finding a publisher.
They were right. Bob took this fish out of water and taught me how to
swim. I owe a debt of gratitude to Tina, Lisa, Dee Dee, and Bob.

Judith Regan was practically bouncing off the walls with energy and
excitement when I first spoke with her about my ideas. Her passion for
publishing this book matched my passion for writing it. And it was great
working with Cassie Jones, Cal Morgan, Bridie Clark—our combined energy and drive made me feel like I was on another campaign. (I guess
maybe I was.) My thanks to everyone at ReganBooks for helping me get
it right.

The gifted and patient members of Team Trippi—David Bender,
Andrew Rossmeissl, and Kathy Lash—read the manuscript through several drafts, helped with research, helped with edits, reminded me of things

I had forgotten—and helped me forget things that are better left forgotten. You were, each in your own way and together, invaluable. I was especially fortunate to have the perspective of a talented writer like David around the house—almost as fortunate as I am to have him as a friend.

And then there is "the very talented" Jess Walter (I prefer to call him "brother Walter"; Kathy calls him her new kitchen), who left his family and moved to the farm to work day and night helping me find the right structure for what I wanted to say. The man was tireless—and when I say brother, I mean it. We collaborated like we'd been doing it forever, completing each other sentences around the breakfast table and poking each other with the reprise, "you don't get it." Jess got it. Thanks also to his researcher, Eric Albrecht. I would plug Jess' own books here, but he won't let me—so Google him.

I also want to acknowledge and thank Tammy Haddad at MSNBC for becoming my guardian angel, encouraging me to follow through and write this thing and, along with Phil Griffin and Rick Kaplan at MSNBC, giving me the time and understanding to finish it. If I am wrong and the revolution *is* televised—these will be the people who put it on the air.

Senator Edward Kennedy ran for president in 1980, and in doing so altered the course of my life in ways I never could have imagined. He once quoted the words of Tennyson, which have special meaning to me now: "I am a part of all who I have met . . ."

Marc Cobb, wherever you are, you changed my life. We ran our hearts out on that old square track in high school—and a part of you has been with me running just off my shoulder ever since.

There are three people from my college days at San Jose State that helped shape any success I've had: Terry Christensen, Louie Barozzi, and Shelby Steele.

Carl Wagner, Jon Haber, the late Paul Tully, John Sasso, and Jack Corrigan—all hailing from the Kennedy campaign in 1980—taught me about organizing, politics, passion, and the cause.

Mike Ford hails from the same tribe of brothers that came together for the 1980 election. What can I say? He taught me all that and more. I wanted him to run the Dean campaign—the man *can* run circles around me—but what's more important is that he, more than anyone outside my family, has been there for me in good times and in bad (and there have been plenty of both).

I want to thank Pat Caddell, who suffered through more 2:00 A.M. phone calls from me than just about anyone and was always there with advice and help. Thanks to Katie Tucker, the mother of my three children. We met on the Kennedy campaign, and she became the best time buyer a political media company could have. Although we are divorced, she remains a great friend who always encouraged me and was more than understanding during the thirteen months I worked on the Dean campaign—a period when I wasn't able to spend as much time with our children as I wanted.

Christine, Jim, and Ted, my three amazing kids, were on the campaign trail with me often, but not often enough. Thanks for understanding, and I am glad you finally got to see the crazy stuff your dad does. Please, please put your hands up next time someone asks if you can make a difference.

I've always had the fortune of having groups of organizers around me that made me look better than I was. The Hog Unit for the Mondale campaign in 1984 (they know who they are) was an amazing group of organizers that carried Fritz to victory in the primary states we worked together.

I've had similar good fortune in the tech and Internet world. I have to thank Steven, Peter, and Michael Sprague, Larry Lessig, John Callahan, Deb Nason, David Weinberger, Andy Rappaport, Howard Rheingold, Doc Searles, Alan Chaplin, Bern Galvin, John Hartman, Jock Gill, Ralph Simon, and far too many more to mention here.

The Dean campaign and all that we accomplished was made possible by the ideas and hard work of countless others who came before us: From Gary Hart's brilliant concentric circle organizing strategy to John McCain's first bold attempt to harness the power of the Internet, there are staff members and candidates who plowed the terrain and helped create what we were able to build.

Moveon.org is another pioneer whose contribution leaves me speechless. Wes Boyd, Joan Blades, Eli Pariser, Zack Exley, and the rest of the Moveon organization (including some two million members) started it all—and taught us so much.

There are others, web communities and blogs too many to mention, like Snackman, BigTim, Ecommerceman, Wildman, and Yayapapdoc from the old bulletin boards of THQ and Wave Systems. The people behind those screen names taught me so much about web communities.

CD Marine, PattyfromVt, and 600,000 people like them from the Dean campaign worked to change our country and succeeded, I believe, in changing politics and steering it the direction of meaning something again. You woke up the Democratic Party and made it the party of the people again.

To the incomparable staff and team of consultants for Dean for America, who worked all those long hours and suffered and muttered under their breath about my "midnight rule"—I wish I could mention all four hundred of you here. You did something that sixty-three losing Democratic campaigns failed to do. You changed everything. You left your mark, and no one will ever be able to take that away. None of what is in these pages would be possible without your efforts. I hope you're all out there enjoying the sunset. To the 600,000 bloggers, volunteers, Meetuppers and supporters of Dean for America, thank you for helping an old warhorse believe again. Joltin' Joe is back because of you, and loaded up with Diet Pepsi for the journey.

Howard Dean didn't always understand me, and I didn't always understand him, but I am certain of this: He wanted to be president of the United States not out of some personal ambition, but because he believed in the responsibility we have as Americans to work for the common good: it's not what's in it for you—it's what's in it for all of us, for our children and theirs. Howard Dean took that responsibility seriously and gave it voice for the first time in decades. He created a wave and a ripple of energy that will wash across our nation for decades to come. Whether you voted for him or not, the Democratic Party and our nation should be grateful for what he and the Dean campaign accomplished. I know I am.

And finally, I owe the deepest thanks to my wife Kathy, who spends too many days unruffling the feathers I ruffle, making smooth all the things that I make rough. She is a saint who knows me better than anyone, and still, for some inexplicable reason, loves and encourages me. She knew I had to do the Dean campaign when I didn't know it for myself. Every day she finds some new way to inspire and impress me.

Thank you all.

Note: For all those other people who helped me along the way (like Jude Barry and Jim Baumbach) and do not appear here or in the book, I apologize. I swear I got you in but the frickin' editors deleted you. What a country.

ACKNOWLEDGMENTS
TO THE 2008 EDITION

I want to thank John and Elizabeth Edwards for inspiring me to work in another presidential campaign. Elizabeth's strength and tenacity taught me to look for and find my own inner resolve. And John Edwards, the rock, who moved the party and the nation further than he will know, renewed my fighting spirit—I am grateful and I am proud to have worked for him and with the rest of the Edwards team. Jay Anania became a close friend, and was an enormous help professionally and personally throughout the 2008 campaign. The Trippi Team—Daren Berringer, Tom Rosmeissl, David Phelps, and Jordan Pietzsch—made their impact on the Edwards campaign for the better and helped me immensely. Jordan still worries that the Clintons will never forgive him for the Politics of Parsing video. Don't worry, Jordan—they won't.

Tammy Haddad has remained a friend and mentor throughout these years after the Dean campaign, as has Lisa DePaulo—thank you both.

In both the first edition of this book and in this new edition, Cal Morgan and Jess Walter gave their advice and friendship, as well as suggesting edits that made this a better effort.

I have had been blessed with friends old and new who have been there for me when I needed them, particularly in the last four years: Jim Tierney, Joe Costello, Jules Radcliff, Ellen Qualls, Amanda Howe, Paul Maslin, John Fairbank, Mame Reiley, Jim Baumbach, Jon Wadsworth, and Paul Blank.

As I jump into another new world as a CBS News political consultant, Katie Couric, Bob Schieffer, and Harry Smith have all been more than patient as I try to find my on-air comfort zone. Jeff Greenfield, Jim Axelrod, Rick Kaplan, Eric Salzman, Carin Pratt, Barbara Fedida Brill, and Paul Friedman, among so many others, have helped show me the ropes, and shared their advice and time. Cedric Grivot has saved me on more occasions than I can count. Thank you all.

The scariest crew are those who have known me the longest. Mike Ford, Peter Goldman, Pat Caddell, David Bender, and John Haber—all were with me before Dean, during Dean, and after Dean. That says as much about their friendship as it does about their sanity.

I owe more than I can say to Congressman Jim Moran, who believed in me and took a chance on me when no one else would.

My three children are not children anymore, but Christine, Jim, and Ted understood one more time why their dad needed to go out and try to change things—and they make me proud every day. And then there is Kathy, my wife and greatest friend, who loves me no matter what crazy mess I get myself (and us) into. I love you all.

INDEX